T0276091

Secondary Fracture Prevention

Secondary Fracture Prevention

An International Perspective

Edited by

Markus J. Seibel

Paul J. Mitchell

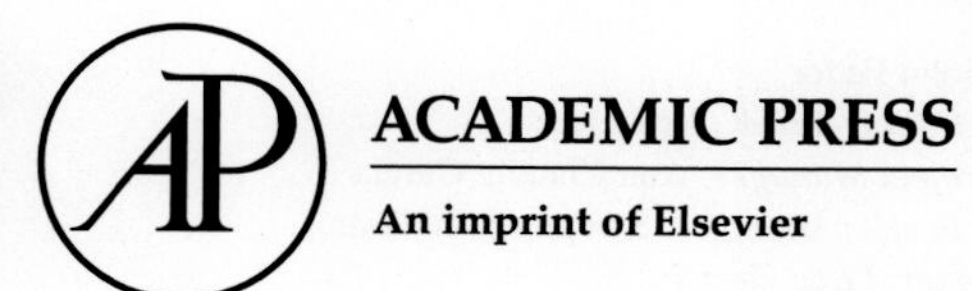

Academic Press is an imprint of Elsevier
125 London Wall, London EC2Y 5AS, United Kingdom
525 B Street, Suite 1650, San Diego, CA 92101, United States
50 Hampshire Street, 5th Floor, Cambridge, MA 02139, United States
The Boulevard, Langford Lane, Kidlington, Oxford OX5 1GB, United Kingdom

Library of Congress Cataloging-in-Publication Data
A catalog record for this book is available from the Library of Congress

British Library Cataloguing-in-Publication Data
A catalogue record for this book is available from the British Library

ISBN: 978-0-12-813136-7

For information on all Academic Press publications visit our website at https://www.elsevier.com/books-and-journals

Publisher: John Fedor
Acquisition Editor: Tari K. Broderick
Editorial Project Manager: Ana Claudia Garcia
Production Project Manager: Swapna Srinivasan
Cover Designer: Greg Harris

Typeset by TNQ Technologies

Contents

List of Contributors xi

1. **Introduction: Why Secondary Fracture Prevention?**
Paul J. Mitchell and Markus J. Seibel

The Ageing Population and Fragility Fractures 1
A Systematic Approach to Fragility Fracture Prevention 2
References 5

2. **The Risk of Osteoporotic Refracture**
Dana Bliuc and Jacqueline R. Center

Introduction 9
Incidence of Osteoporotic Fracture 9
Incidence of Osteoporotic Fracture According to Type of Fracture 10
Secular Trends for Initial Fracture Risk 11
Risk of Subsequent Fracture Following Osteoporotic Fracture 12
The Role of Type of Initial Fracture in Subsequent Fracture Risk 13
Timing of Subsequent Fracture 15
Role of BMD in Predicting Subsequent Fracture Risk 16
Secular Trends for Subsequent Fracture Risk 17
Risk of Mortality Following Osteoporotic Fractures 18
Mortality Risk According to Fracture Type 18
The Role of Time Post Initial Fracture 20
Subsequent Fracture and Mortality Risk 21
Secular Trends in Postfracture Mortality Risk 21
The Association Between Antiresorptive Medication and Mortality Risk Reduction 22
Conclusion 23
References 24

3. **Models of Secondary Fracture Prevention: Systematic Review and Metaanalysis of Outcomes**
Kirtan Ganda, Paul J. Mitchell and Markus J. Seibel

Secondary Fracture Prevention Programmes 33
Search Methodology 34

Type A Model Interventions 35
Type B Model Interventions 43
Type C Model Interventions 48
Type D Model Interventions 53
Summary and Conclusions 57
References 59

4. Fracture Liaison Services: An Australasian Perspective
Jacqueline C.T. Close

Introduction 63
Australia 63
New Zealand 64
The Osteoporosis Care Gap 65
Closing the Osteoporosis Care Gap 65
New Zealand 66
Australia 67
Hip Fracture Care – United We Stand 70
ANZ Hip Fracture Registry 70
Summary 71
Acknowledgements 74
References 74

5. Fracture Liaison Services – Canada
Victoria Elliot-Gibson, Joanna Sale, Ravi Jain and Earl Bogoch

Introduction 79
Canadian Health Care System 80
Access to Bone Mineral Density Testing in Canada 80
Access to Pharmacotherapy for Fracture Prevention in Canada 81
Fracture Risk Assessment Tools in Canada 81
The Role of Osteoporosis Canada in Supporting the Implementation of Quality FLS in Canada 81
Ontario Provincial Fracture Liaison Service 83
Ontario Wide Fracture Screening and Prevention Programme 83
Nova Scotia Provincial Fracture Liaison Services 86
Alberta Provincial Fracture Liaison Services 87
Independent Fracture Liaison Services in Canada 88
Fracture Liaison Services at St. Michael's Hospital, Toronto, Ontario, Canada 88
Cost-Effectiveness of Fracture Liaison Services in Canada 96
The Patient Experience: Making Fracture Liaison Services Work Better 97
Patients Do Not Associate Their Fragility Fracture With Their Underlying Bone Health 97
Burden of Fracture 98

Education Is Not Enough 98
Pharmacotherapy Use: Uptake and Persistence 99
Reliance on the Concept of 'Fracture Risk' May Be Problematic 99
Barriers at the Health-Care Provider Level 100
Conclusions 100
The Future of Fracture Liaison Services in Canada 102
Acknowledgements 102
References 103

6. The Challenge of Secondary Prevention of Hip Fracture in Japan

Noriaki Yamamoto, Hideaki E. Takahashi and Naoto Endo

Background 109
Hip Fracture Clinical Pathway in Japan 109
Osteoporosis Liaison Service and Osteoporosis Manager 110
Fragility Fracture Network Japan 110
National Hip Fracture Database Japan Project 111
Projects in Niigata Prefecture 112
Conclusion 115
References 115

7. Secondary Fracture Prevention: Lebanon

Ghassan Maalouf and Maroun Rizkallah

Introduction 117
Epidemiology of Fragility Fractures in Lebanon 117
Scientific Contribution 118
Fracture Liaison Service 119
Contributions by the Health Authorities 119
Conclusion 120
References 120

8. Fracture Liaison Services in South East Asia: Notes From a Large Public Hospital in Singapore

Manju Chandran

Introduction 123
The Singapore Story 123
Strengths of Osteoporosis Patient Targeted and Integrated Management for Active Living 129
Where Have We 'Dropped the Ball'? 129
Where Do We Go From Here? 130
Conclusion 131
Acknowledgements 131
References 131

9. **Fracture Liaison Services in Taiwan: Developments and Perspectives**
Lo-Yu Chang and Ding-Cheng (Derrick) Chan

Introduction 133
The Development of Osteoporosis Care Management Services in Taiwan 134
The National Taiwan University Hospital Model 134
Promotion of the Fracture Liaison Service in Taiwan 136
Fracture Liaison Services in Taiwan: Preliminary Outcomes 142
Challenges in Establishing Fracture Liaison Services in Taiwan 143
References 143

10. **International Models of Secondary Fracture Prevention: United Kingdom**
Muhammad Kassim Javaid and Paul J. Mitchell

The UK National Health Service 145
It Started in Glasgow, Scotland 146
The UK Top-Down Approach 146
The UK Bottom-Up Approach 149
Current Challenges – Delivering for Every Patient 150
Acknowledgments 151
Disclosure 151
References 152

11. **Fracture Liaison Service: US Perspective**
Thomas P. Olenginski

Introduction 155
Kaiser Permanente Healthy Bones Programme 157
Geisinger High-Risk Osteoporosis Clinic Programme 160
Fracture Liaison Service Demonstration Project 165
Own the Bone 166
Registries/Databases 167
Quality Reporting 168
Conclusions 169
Acknowledgements 169
References 170

12. **National and International Programs**
Paul J. Mitchell

Australia 173
FLS/SFPP Implementation Initiatives 173
Clinical Standards 173
Fracture Registries 174

Canada 174
FLS/SFPP Implementation Initiatives 174
Clinical Standards 175
Fracture Registries 175
New Zealand 175
FLS/SFPP Implementation Initiatives 175
Clinical Standards 176
Fracture Registries 177
United Kingdom 178
FLS/SFPP Implementation Initiatives 178
Clinical Standards 178
Fracture Registries 179
United States of America 180
Clinical Standards 180
Fracture Registries 180
International Initiatives 180
International Osteoporosis Foundation Capture the Fracture Program 180
Fragility Fracture Network 181
References 182

Appendix Summary of Useful Resources 185
Index 187

List of Contributors

Dana Bliuc Osteoporosis and Bone Biology Program, Garvan Institute of Medical Research, Sydney, NSW, Australia

Earl Bogoch Mobility Program. St. Michael's Hospital, Toronto, ON, Canada; Li Ka Shing Knowledge Institute, St. Michael's Hospital, Toronto, ON, Canada; Department of Surgery, University of Toronto, Toronto, ON, Canada

Jacqueline R. Center Osteoporosis and Bone Biology Program, Garvan Institute of Medical Research, Sydney, NSW, Australia; Clinical School, St Vincent's Hospital, Sydney, NSW, Australia; Faculty of Medicine, UNSW, Sydney, NSW, Australia

Ding-Cheng (Derrick) Chan Department of Geriatrics and Gerontology, National Taiwan University Hospital, Taipei, Taiwan; Department of Internal Medicine, National Taiwan University Hospital, Taipei, Taiwan; Superintendent Office, Chu-Tung Branch, National Taiwan University Hospital, Hsinchu County, Taiwan

Manju Chandran Osteoporosis and Bone Metabolism Unit, Department of Endocrinology, Singapore General Hospital, Singapore

Lo-Yu Chang School of Medicine, National Taiwan University, Taipei, Taiwan

Jacqueline C.T. Close Neuroscience Research Australia and Prince of Wales Hospital Clinical School, University of New South Wales, Sydney, Australia

Victoria Elliot-Gibson Musculoskeletal Health and Outcomes Research, Li Ka Shing Knowledge Institute, St. Michael's Hospital, Toronto, ON, Canada; Mobility Program. St. Michael's Hospital, Toronto, ON, Canada; Ontario Osteoporosis Strategy, Osteoporosis Canada, Toronto, ON, Canada

Naoto Endo Department of Orthopedic Surgery, Niigata University School of medicine, Niigata, Japan

Kirtan Ganda Department of Endocrinology and Metabolism, Concord Repatriation General Hospital, Concord, NSW, Australia; Concord Clinical School, University of Sydney, Sydney, NSW, Australia

Ravi Jain Ontario Osteoporosis Strategy, Osteoporosis Canada, Toronto, ON, Canada

Muhammad Kassim Javaid National Institute of Health Research Oxford Biomedical Research Centre, Nuffield Department of Orthopaedics, Rheumatology and Musculoskeletal Sciences, University of Oxford, Oxford, United Kingdom

Ghassan Maalouf Orthopedic Surgery Department, Bellevue University Medical Center, Saint-Joseph University, Beirut, Lebanon

Paul J. Mitchell School of Medicine, Sydney Campus, University of Notre Dame Australia, Sydney, NSW, Australia; Osteoporosis New Zealand, Wellington, New Zealand

Thomas P. Olenginski Department of Rheumatology, Geisinger Health System, Danville, PA, United States

Maroun Rizkallah Orthopedic Surgery Department, Bellevue University Medical Center, Saint-Joseph University, Beirut, Lebanon

Joanna Sale Musculoskeletal Health and Outcomes Research, Li Ka Shing Knowledge Institute, St. Michael's Hospital, Toronto, ON, Canada; Institute of Health Policy, Management & Evaluation, University of Toronto, Toronto, ON, Canada; Li Ka Shing Knowledge Institute, St. Michael's Hospital, Toronto, ON, Canada

Markus J. Seibel Department of Endocrinology and Metabolism, Concord Repatriation General Hospital, Concord, NSW, Australia; Concord Clinical School, University of Sydney, Sydney, NSW, Australia; Bone Research Program, ANZAC Research Institute, Sydney, NSW, Australia

Hideaki E. Takahashi Department of Orthopedic Surgery, Niigata Rehabilitation Hospital, Niigata, Japan

Noriaki Yamamoto Department of Orthopedic Surgery, Niigata Rehabilitation Hospital, Niigata, Japan

Chapter 1

Introduction: Why Secondary Fracture Prevention?

Paul J. Mitchell[1,2], **Markus J. Seibel**[3,4,5]
[1]School of Medicine, Sydney Campus, University of Notre Dame Australia, Sydney, NSW, Australia; [2]Osteoporosis New Zealand, Wellington, New Zealand; [3]Department of Endocrinology and Metabolism, Concord Repatriation General Hospital, Concord, NSW, Australia; [4]Concord Clinical School, University of Sydney, Sydney, NSW, Australia; [5]Bone Research Program, ANZAC Research Institute, Sydney, NSW, Australia

THE AGEING POPULATION AND FRAGILITY FRACTURES

The global population is ageing fast. In 2015, the Department of Economic and Social Affairs of the United Nations estimated that between 2015 and 2030, the number of people aged 60 years or over would increase by 56%, from 901 million to more than 1.4 billion.[1] The number of individuals classified as the 'oldest old'—people aged 80 years and over—was projected to increase from 125 million in 2015 to 202 million and 434 million by 2030 and 2050, respectively. A consequence of this unprecedented shift in the demographic composition of our society will be a dramatic rise in the number of people living with chronic, age-related conditions. As noted in the 2014 World Osteoporosis Day Report, 'osteoporosis will be at the vanguard of this battle set to rage between quantity and quality of life'.[2]

At the beginning of this millennium, the annual global incidence of osteoporotic fragility fractures was estimated to be 9 million, of which 1.6 million were at the hip, 1.7 million at the forearm, 1.4 million were clinical vertebral fractures,[a] 0.7 million at the humerus, and 3.6 million fractures at other sites, resulting in 5.8 million disability-adjusted life-years lost.[3] In 2010, the number of individuals aged 50 years or over who were at high risk of osteoporotic fracture worldwide was estimated to be 158 million, a figure which is set to double by 2040.[4]

Chapter 2 of this book provides a comprehensive overview of the epidemiology of fragility fractures, with a particular focus on secondary fractures.

---

a. As this number relates to clinical (i.e., symptomatic) vertebral fractures only, and as 60% of all vertebral fractures are initially asymptomatic, the true incidence of vertebral fractures is likely to be significantly higher.

Secondary Fracture Prevention. https://doi.org/10.1016/B978-0-12-813136-7.00001-6

A SYSTEMATIC APPROACH TO FRAGILITY FRACTURE PREVENTION

Faced with an impending tsunami of fragility fractures, where should policymakers, healthcare professionals, payers and patients start to tackle this problem? To make the maximum impact in the shortest possible time frame, a pragmatic approach would be to target bone health assessments at individuals who are at highest risk of sustaining fragility fractures and for whom current treatment strategies have been shown to be most effective in reducing fracture risk. In this regard, individuals who have already sustained a fragility fracture are the most obvious population to address first.

Fracture begets fracture. Since the 1980s, we have known that approximately half of individuals who sustain a hip fracture broke another bone in the months or years before breaking their hip.[5] More recently, investigators in Australia,[6] the United Kingdom[7] and the United States[8] have reported similar findings. Metaanalyses have demonstrated that a prior fracture at any site is associated with a doubling of future fracture risk.[9,10] Further, there is evidence that fractures cluster in time. An analysis of individuals managed by the Glasgow Fracture Liaison Service (FLS) in Scotland found that 80% of refractures occurred during the first year after the index fracture.[11] Indeed, half of refractures occurred within 6–8 months, dependent on whether the incident fracture was hip (6 months) or non-hip (8 months).

Targeting individuals who have sustained fragility fractures is also a manageable undertaking. Epidemiological studies from France,[12] Germany,[13] Italy,[14] Sweden[15] and the United Kingdom[16] reported that among women aged 50 years or over, between 10.4% (France) and 22.8% (Sweden) had a history of fragility fracture. Taking into account the retrospective analyses of prior fracture history among individuals who sustain hip fractures, as described above, these data would suggest that approximately half of hip fractures emanate from a 10th to a quarter of the female population aged 50 years or over.

In 2017, Harvey et al. published a narrative review focused on current and future strategies for prevention of fragility fractures.[17] The authors stated that 'secondary fracture prevention is an obvious first step in the development of a systematic approach to prevention of all fragility fractures caused by osteoporosis'. Indeed, a broad range of agents have been shown to reduce the incidence of subsequent (secondary) fractures in individuals who have sustained a prior fragility fracture by up to 70%.[18–25]

Individuals who sustain fragility fractures usually present themselves to healthcare services to seek medical attention. The institutions to which fracture patients present themselves may vary between and within countries. A recent study from New Zealand sought to determine how many fragility fractures occurred in the central Auckland region during the period 2011–12, and what proportion of the individuals who sustained fractures received secondary preventive care.[26] Of the 2729 fracture patients, 56% were seen at Auckland City

Hospital, whereas the remaining 44% were managed in the private outpatient sector. Notwithstanding the organisational challenges that such variation in care-seeking behaviour can present to healthcare professionals, the majority of patients with nonvertebral fragility fractures are readily identifiable. Unfortunately, this is not the case for a significant proportion of vertebral fractures, which often go unrecognised, undiagnosed and untreated. Identification of these patients requires innovative methods such as natural language processing of radiology reports[27] or automated reading and analysis of spinal imaging studies.

During the last 15 years, a considerable amount of activity has been undertaken to develop systematic approaches to secondary fracture prevention. The approach taken in New Zealand,[28] as illustrated in Fig. 1.1, is representative of similar approaches implemented in several countries, including Australia,[29] Canada,[30] the United Kingdom[31] and the United States.[32] The top two priorities in this approach relate to secondary fracture prevention.

Individuals who sustain a hip fracture are at significant risk of sustaining a second hip fracture.[33–36] To address this issue and support improvements in acute care of hip fracture patients, the Fragility Fracture Network (http://fragilityfracturenetwork.org/) has advocated for development of consensus guidelines, quality standards, and systematic performance measurement.[37] This approach has been taken in several countries, including Australia,[38–40] New Zealand[38–40] and the United Kingdom.[41,42] National hip fracture registries provide a mechanism for benchmarking of quality standards[43,44] provided by hospitals. In the United Kingdom significant improvements have been made during the decade since the launch of the National Hip Fracture Database (NHFD).[45,46] In 2016, Rath and colleagues compared specific measures of hip fracture care and secondary

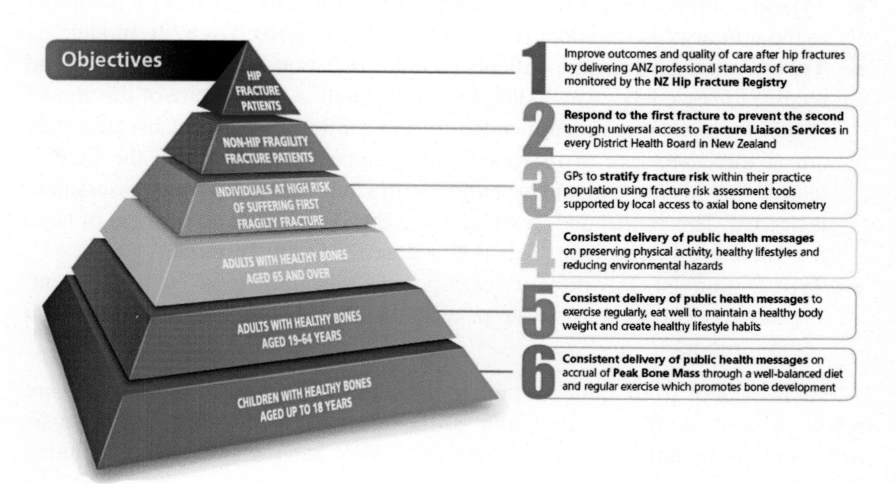

FIGURE 1.1 A systematic approach to fragility fracture care and prevention.[28] *(Reproduced with kind permission of Osteoporosis New Zealand.)*

prevention for hip fracture patients admitted to a Beijing tertiary hospital in 2012 (n = 780) against the findings of the UK NHFD for hip fracture patients presenting to 180 UK hospitals in the same year (n = 59,365).[47] Key findings included:

- Osteoporosis assessment: 0.3% of patients in Beijing versus 94% in the United Kingdom.
- Falls assessment: 3.8% of patients in Beijing versus 92% in the United Kingdom.

The authors concluded that development of a China Hip Fracture Registry, in combination with financial incentives for hospitals to deliver best practice (as is done in the United Kingdom), has the potential to transform the management of hip fracture in China.

An FLS, also known as a Secondary Fracture Prevention Program (SFPP), is a system to ensure fracture risk assessment in, and treatment as appropriate of all patients presenting with one or more fragility fractures. An FLS/SFPP usually comprises a dedicated case worker, often a clinical nurse specialist or young physician, who works to pre-agreed protocols to case-find and assess fracture patients.

In 2013, Ganda et al. published a systematic review and metaanalysis of the outcomes achieved by FLS/SFPP.[48] A useful classification system for models of care was devised according to the intensity of service provision:

- **Type A models:** Identifies, investigates and initiates treatment, where appropriate, for fragility fracture patients. This is also described as a '3i' model.
- **Type B models:** Identifies and investigates but leaves the initiation of treatment to the primary care provider (PCP). This is also described as a '2i' model.
- **Type C models:** Fracture patients receive education about osteoporosis and receive lifestyle advice including falls prevention. A key feature of this model is that the patient is recommended to seek further assessment because they are at increased risk of osteoporosis and repeat fractures, and the PCP is alerted that his/her patient has suffered a fracture and that further assessment is needed. This model does not undertake BMD testing or assessment of need for osteoporosis treatment. This is also described as a '1i' model.
- **Type D models:** Only provides osteoporosis education to the fracture patient. Type D models do not educate or alert the PCP. This is also described as a 'Zero i' model.

As is evident from Table 1.1, the more intensive FLS/SFPP models of care result in more patients undergoing BMD testing and receiving osteoporosis treatment. An update to Ganda's systematic review and metaanalysis is included in Chapter 3.

Further, effective FLS/SFPPs have been established in both the primary care and secondary care settings. Although it is the case that the majority of publications relating to the more intensive and, therefore, effective FLS/SFPP

TABLE 1.1 FLS/SFPP Models of Care of Varying Intensity and Outcomes[48]

FLS/SFPP Model	Proportion Investigated With BMD Testing	Proportion Initiated on Osteoporosis Treatment
Type A: 3i FLS/SFPP model	79%	46%
Type B: 2i FLS/SFPP model	60%	41%
Type C: 1i FLS/SFPP model	43%	23%
Type D: 'Zero i' FLS/SFPP model	–	8%

Reproduced with kind permission of Springer.

(i.e., Type A or B models) describe programs based in secondary care, high-performing primary care-based programs have been established in the United Kingdom.[49,50] These programs illustrate that the setting does not determine success or failure. The organisation and staffing of the program is crucial.

The funding situation in the UK NHS makes it increasingly likely that successful and sustainable FLS/SFPP will be funded by primary care and, therefore, commonly be located in primary care. The Commonwealth–state divide of funding streams in Australia is highly analogous to the UK situation. Accordingly, Primary Health Networks should be candidate organisations in the Australian public health system to fund FLS/SFPP in the future. Appropriately staffed and funded Type A or B FLS/SFPP ensure that the majority of fragility fracture patients receive the postfracture care that they need.

REFERENCES

1. United Nations Department of Economic and Social Affairs Population Division. *World population ageing 2015-highlights (ST/ESA/SER.A/368)*. New York. 2015.
2. Ebeling PR. *Osteoporosis in men: why change needs to happen*. Nyon.: International Osteoporosis Foundation; 2014.
3. Johnell O, Kanis JA. An estimate of the worldwide prevalence and disability associated with osteoporotic fractures. *Osteoporos Int* 2006;**17**(12):1726–33.
4. Oden A, McCloskey EV, Kanis JA, Harvey NC, Johansson H. Burden of high fracture probability worldwide: secular increases 2010-2040. *Osteoporos Int* 2015;**26**(9):2243–8.
5. Gallagher JC, Melton LJ, Riggs BL, Bergstrath E. Epidemiology of fractures of the proximal femur in Rochester, Minnesota. *Clin Orthop Relat Res* 1980;(150):163–71.
6. Port L, Center J, Briffa NK, Nguyen T, Cumming R, Eisman J. Osteoporotic fracture: missed opportunity for intervention. *Osteoporos Int* 2003;**14**(9):780–4.
7. McLellan A, Reid D, Forbes K, et al. *Effectiveness of strategies for the secondary prevention of osteoporotic fractures in Scotland (CEPS 99/03)*. NHS Quality Improvement Scotland; 2004.

8. Edwards BJ, Bunta AD, Simonelli C, Bolander M, Fitzpatrick LA. Prior fractures are common in patients with subsequent hip fractures. *Clin Orthop Relat Res* 2007;**461**:226–30.
9. Klotzbuecher CM, Ross PD, Landsman PB, Abbott 3rd TA, Berger M. Patients with prior fractures have an increased risk of future fractures: a summary of the literature and statistical synthesis. *J Bone Miner Res* 2000;**15**(4):721–39.
10. Kanis JA, Johnell O, De Laet C, et al. A meta-analysis of previous fracture and subsequent fracture risk. *Bone* 2004;**35**(2):375–82.
11. Langridge CR, McQuillian C, Watson WS, Walker B, Mitchell L, Gallacher SJ. Refracture following fracture liaison service assessment illustrates the requirement for integrated falls and fracture services. *Calcif Tissue Int* 2007;**81**(2):85–91.
12. Cawston H, Maravic M, Fardellone P, et al. Epidemiological burden of postmenopausal osteoporosis in France from 2010 to 2020: estimations from a disease model. *Arch Osteoporos* 2012;**7**:237–46.
13. Gauthier A, Kanis JA, Jiang Y, et al. Burden of postmenopausal osteoporosis in Germany: estimations from a disease model. *Arch Osteoporos* 2012;**7**:209–18.
14. Piscitelli P, Brandi M, Cawston H, et al. Epidemiological burden of postmenopausal osteoporosis in Italy from 2010 to 2020: estimations from a disease model. *Calcif Tissue Int* 2014;**95**(5):419–27.
15. Gauthier A, Kanis JA, Martin M, et al. Development and validation of a disease model for postmenopausal osteoporosis. *Osteoporos Int* 2011;**22**(3):771–80.
16. Gauthier A, Kanis JA, Jiang Y, et al. Epidemiological burden of postmenopausal osteoporosis in the UK from 2010 to 2021: estimations from a disease model. *Arch Osteoporos* 2011;**6**:179–88.
17. Harvey NC, McCloskey EV, Mitchell PJ, et al. Mind the (treatment) gap: a global perspective on current and future strategies for prevention of fragility fractures. *Osteoporos Int* 2017;**28**(5):1507–29.
18. Wells GA, Cranney A, Peterson J, et al. Alendronate for the primary and secondary prevention of osteoporotic fractures in postmenopausal women. *Cochrane Database Syst Rev* 2008;(1):CD001155.
19. Wells GA, Cranney A, Peterson J, et al. Etidronate for the primary and secondary prevention of osteoporotic fractures in postmenopausal women. *Cochrane Database Syst Rev* 2008;(1):CD003376.
20. Wells G, Cranney A, Peterson J, et al. Risedronate for the primary and secondary prevention of osteoporotic fractures in postmenopausal women. *Cochrane Database Syst Rev* 2008;(1):CD004523.
21. Palacios S, Kalouche-Khalil L, Rizzoli R, et al. Treatment with denosumab reduces secondary fracture risk in women with postmenopausal osteoporosis. *Climacteric* 2015;**18**(6):805–12.
22. Ettinger B, Black DM, Mitlak BH, et al. Reduction of vertebral fracture risk in postmenopausal women with osteoporosis treated with raloxifene: results from a 3-year randomized clinical trial. Multiple Outcomes of Raloxifene Evaluation (MORE) Investigators. *JAMA* 1999;**282**(7):637–45.
23. Neer RM, Arnaud CD, Zanchetta JR, et al. Effect of parathyroid hormone (1-34) on fractures and bone mineral density in postmenopausal women with osteoporosis. *N Engl J Med* 2001;**344**(19):1434–41.
24. Lyles KW, Colon-Emeric CS, Magaziner JS, et al. Zoledronic acid and clinical fractures and mortality after hip fracture. *N Engl J Med* 2007;**357**(18):1799–809.
25. Eastell R, Black DM, Boonen S, et al. Effect of once-yearly zoledronic acid five milligrams on fracture risk and change in femoral neck bone mineral density. *J Clin Endocrinol Metab* 2009;**94**(9):3215–25.
26. Braatvedt G, Wilkinson S, Scott M, Mitchell P, Harris R. Fragility fractures at Auckland City Hospital: we can do better. *Arch Osteoporos* 2017;**12**(1):64.

27. Pandya J, Ganda K, Seibel MJ. *Systems based Identification of patients with osteoporotic vertebral fractures* (in press). 2018.
28. Osteoporosis New Zealand. *Osteoporosis New Zealand strategic plan 2017 – 2020*. 2017. https://osteoporosis.org.nz/about-us/our-strategy/.
29. New South Wales Agency for Clinical Innovation Musculoskeletal Network. In: *NSW model of care for osteoporotic refracture prevention*. 2011. Chatswood, NSW.
30. Osteoporosis Canada. *Make the first break the last with fracture Liaison services*. 2013.
31. Department of Health. In: *Falls and fractures: effective interventions in health and social care*. Department of Health; 2009.
32. Cooper C, Dawson-Hughes B, Gordon CM, Rizzoli R. *Healthy nutrition, healthy bones: how nutritional factors affect musculoskeletal health throughout life*. Nyon.: International Osteoporosis Foundation; 2015.
33. Lawrence TM, Wenn R, Boulton CT, Moran CG. Age-specific incidence of first and second fractures of the hip. *J Bone Joint Surg Br* 2010;**92**(2):258–61.
34. Fukushima T, Sudo A, Uchida A. Bilateral hip fractures. *J Orthop Sci* 2006;**11**(5):435–8.
35. Lee YK, Ha YC, Yoon BH, Koo KH. Incidence of second hip fracture and compliant use of bisphosphonate. *Osteoporos Int* 2013;**24**(7):2099–104.
36. Chen FP, Shyu YC, Fu TS, et al. Secular trends in incidence and recurrence rates of hip fracture: a nationwide population-based study. *Osteoporos Int* 2017;**28**(3):811–8.
37. Fragility Fracture Network. *Fragility fracture network website*. 2017. http://fragilityfracturenetwork.org/.
38. Australian and New Zealand Hip Fracture Registry. *Australian and New Zealand hip fracture registry website*. 2015. http://www.anzhfr.org/.
39. Australian and New Zealand Hip Fracture Registry. *2016 annual report*. Sydney. 2016.
40. Australian and New Zealand Hip Fracture Registry. *ANZ hip fracture Registry newsletter June 2017*. Sydney. 2017.
41. Royal College of Physicians. *National hip fracture database (NHFD) annual report 2016*. London: RCP; 2016.
42. Royal College of Physicians. *The national hip fracture database*. 2017. http://www.nhfd.co.uk/.
43. National Institute for Health and Clinical Excellence. *Quality standard for hip fracture care*. NICE Quality Standard 16. London. 2012.
44. Australian Commission on Safety and Quality in Health Care. *Health quality & safety commission New Zealand*. Hip Fracture Care Clinical Care Standard. Sydney. 2016.
45. Neuburger J, Currie C, Wakeman R, et al. The impact of a national clinician-led audit initiative on care and mortality after hip fracture in England: an external evaluation using time trends in non-audit data. *Med Care* 2015;**53**(8):686–91.
46. Wise J. Hip fracture audit may have saved 1000 lives since 2007. *BMJ* 2015;**351**:h3854.
47. Tian M, Gong X, Rath S, et al. Management of hip fractures in older people in Beijing: a retrospective audit and comparison with evidence-based guidelines and practice in the UK. *Osteoporos Int* 2016;**27**(2):677–81.
48. Ganda K, Puech M, Chen JS, et al. Models of care for the secondary prevention of osteoporotic fractures: a systematic review and meta-analysis. *Osteoporos Int* 2013;**24**(2):393–406.
49. Brankin E, Mitchell C, Munro R, Lanarkshire Osteoporosis S. Closing the osteoporosis management gap in primary care: a secondary prevention of fracture programme. *Curr Med Res Opin* 2005;**21**(4):475–82.
50. Chan T, de Lusignan S, Cooper A, Elliott M. Improving osteoporosis management in primary care: an audit of the impact of a community based fracture liaison nurse. *PLoS One* 2015;**10**(8):e0132146.

Chapter 2

The Risk of Osteoporotic Refracture

Dana Bliuc[1], Jacqueline R. Center[1,2,3]
[1]Osteoporosis and Bone Biology Program, Garvan Institute of Medical Research, Sydney, NSW, Australia; [2]Clinical School, St Vincent's Hospital, Sydney, NSW, Australia; [3]Faculty of Medicine, UNSW, Sydney, NSW, Australia

INTRODUCTION

Osteoporosis is a systemic skeletal disorder characterised by reduced bone mass and microarchitectural deterioration of bone tissue, which results in an increased risk of fracture.[1] Osteoporosis is a major public health burden which has been estimated to cause over 8.9 million osteoporotic fractures annually worldwide.[2] It is particularly common in industrialised countries such as North America, Japan, Western Europe, Australia and New Zealand, and it is predicted to increase for the developing regions. Osteoporotic fractures are associated with significant direct and indirect costs, estimated at 32 billion EUR per year in Europe,[3] 17 billion USD per year in the United States[4] and 1.9 billion AUD per year in Australia.[5] In addition, osteoporotic fractures lead to disability, further fracture and premature mortality. This burden is further aggravated by the lack of appropriate management of patients with osteoporosis postfracture despite the availability of highly effective therapies.

INCIDENCE OF OSTEOPOROTIC FRACTURE

The incidence of osteoporotic fractures in the community increases significantly after the age of 50, for both women and men. The pattern of fracture incidence depends largely on the type of fracture. Hip and vertebral fracture incidence increases exponentially with age, whereas nonhip nonvertebral fractures (NHNV) have a much higher incidence in the younger age groups with less of an age-related increase (Fig. 2.1).

Secondary Fracture Prevention. https://doi.org/10.1016/B978-0-12-813136-7.00002-8

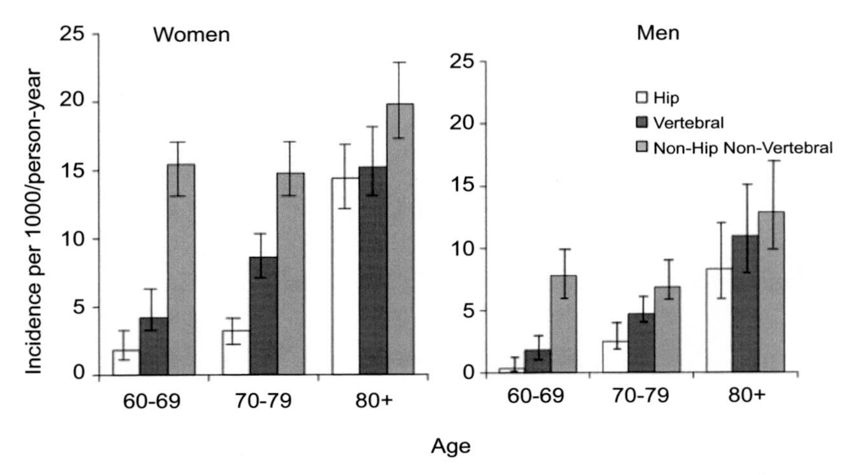

FIGURE 2.1 Cumulative incidence of initial fracture type in Dubbo population according to age of women and men.[6]

Incidence of Osteoporotic Fracture According to Type of Fracture

Hip Fracture

The mean age of hip fracture is ~80 years for women and 76 for men.[7] The risk of hip fracture is higher for women with the overall female: male ratio estimated ~2:1.[8] Despite the increasing risk of hip fracture with age, the majority of hip fractures still occurred before the age of 80 in both women and men (Fig. 2.2).

The incidence of hip fracture varies widely across the globe. In a recent metaanalysis using data from 63 countries, the highest age-standardized incidence of hip fracture was observed in Northern and Central European countries such as Norway, Sweden, Ireland, Denmark, Belgium and Austria, whereas the lowest incidence was observed in Nigeria, South Africa, Tunisia and Ecuador.[9]

Vertebral Fracture

Vertebral fractures are the most common fracture type, but among the least well reported. Unlike hip fracture, only 30%–40% of all vertebral fractures are symptomatic and come to medical attention,[10,11] whereas the remainder are asymptomatic and only identified by incidental radiographs.

The prevalence of vertebral fracture ranges between 20% and 25%[12,13] and it is higher in men than in women. The incidence of new vertebral fracture has been assessed in The European Prospective Osteoporosis Study,[14] the Rotterdam study in the Netherlands[15] and the Study of Osteoporotic Fractures in the United States.[16] By contrast with the prevalence data, the incidence of vertebral fracture was higher in women (10.7–14.7 vertebral fractures/1000 person-year) than in men (5.7–5.9 vertebral fractures/1000 person-year). This discrepancy may

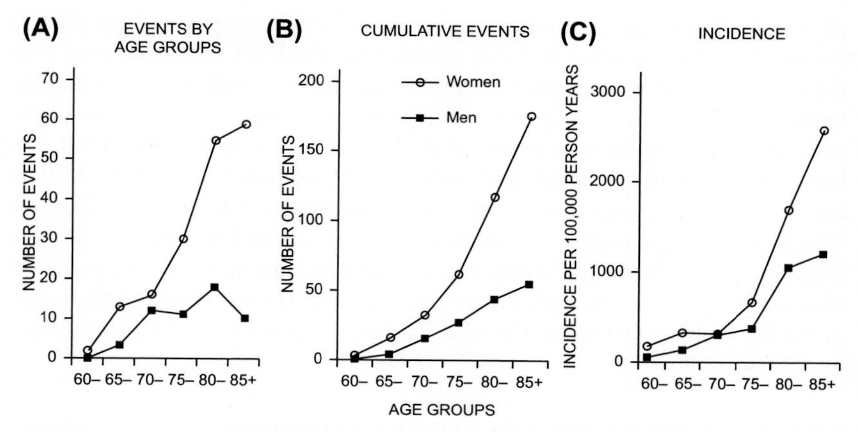

FIGURE 2.2 Hip fracture incidence expressed in (A) absolute number of fractures, (B) cumulative number of fractures and (C) incidence rate per 100,000 person-years. (A) The absolute number of hip fractures in men peaked at the ages of 80–84 and then fell, whereas fractures in women continuously rise with age. (B) Nearly one-half (48%) of hip fractures in men occurred before the age of 80, and 66% of hip fractures in women occurred before the age of 85. (C) The incidence rate increased exponentially with advancing age.[8]

reflect a higher proportion of nonosteoporotic traumatic vertebral fractures in men than women.[17] Additionally, over the age of 50, women presenting with back pain are more likely to have X-rays and to be classified as 'atraumatic fractures' compared to men.[18]

Nonhip Nonvertebral Fracture

NHNVs account for over 40% of all osteoporotic fractures occurring in the community[6] and more than 20% of the total costs.[4] In contrast to hip and vertebral fractures, their incidence is more constant across all ages (Fig. 2.1). For example, in the Dubbo Osteoporosis Epidemiology Study, the incidence of NHNV fracture for the 60–69 year age group was 15.4 (95% CI, 12.7–18.8)/1000 person-years for women and 7.8 (95% CI, 5.8–10.4)/1000 person-years for men compared with a clinical vertebral fracture incidence of 4.2 (95%CI, 2.9–6.1)/1000 person-years for women and 1.8 (95% CI, 1.0–3.3)/1000 person-years for men in the same age group. By 80+ years of age, the incidence of these fracture types is more similar, but the average incidence of the NHNV fractures remains higher than that of hip or vertebral fractures.[6]

Secular Trends for Initial Fracture Risk

There is emerging evidence that the incidence of osteoporotic fractures and, in particular, hip fracture is declining in developed countries.[19,20] It is not clear whether this decline is due to increased medication uptake, or better overall health care. By contrast, studies from Asian countries have suggested either a plateauing or increase in hip fracture incidence.[21,22] There is very little data on the secular trends of other types of osteoporotic fractures. However, even with the decline in the osteoporotic fracture incidence, it is still expected that the number of osteoporotic fractures will increase over time. The segment of the population over 65 years is expected to increase as a result of the significant improvement in life expectancy. Gullberg et al. estimated that the 1.3 million hip fractures calculated using worldwide demographic data for 1990 from the World Health Organisation will increase progressively to 2.6 million by the year 2025 and to 4.5 million by 2050, with a higher increase in men than women.[23] More recently, a study from Norway estimated that even with the current decline in the hip fracture incidence, the total burden of hip fracture will still increase by 22% by 2040.[24]

RISK OF SUBSEQUENT FRACTURE FOLLOWING OSTEOPOROTIC FRACTURE

It has been consistently shown that the occurrence of a fragility fracture after the age of 50 is associated with an increased risk of future fragility/osteoporotic fracture. This risk has been demonstrated not only for the fractures widely recognised as part of the 'osteoporotic fracture triad' (hip, vertebral and forearm)[25–31] but also for all osteoporotic fractures.[30–32] The magnitude of subsequent fracture risk increase at any skeletal site ranges from 1.8- to 2.8-fold depending on the age and gender structure of the population studied.[32–36] Similar to initial fracture risk, the risk of subsequent fracture increases with increasing age. In a study from Tromsø, Norway, the proportion of individuals who sustained a subsequent fracture increased from 9% to 30% in women and from 10% to 26% for men from the age groups of 50–59 to 80+ years. Most importantly, the risk of subsequent fracture was between 1.7- and 2.0-fold higher than the risk of initial fracture for both younger and older individuals (RR, 1.7; 95% CI, 1.4–2.1 and 1.3; 95% CI, 1.0–1.6 for women and 2.0; 95% CI, 1.6–2.4 and 1.4 (0.9–2.3) for men of 50–59 and 80+ years of age).[30] A similarly increased absolute risk of subsequent fracture with age has been reported in several other cohorts.[32,37]

There are far fewer studies on the risk of subsequent fractures conducted in men than women.[27] However, epidemiological cohort studies that provided separate estimates of subsequent fracture for each gender have shown that men, despite having a significantly lower absolute risk of initial fracture than women, have a similar risk of subsequent fracture. In the Dubbo Osteoporosis Epidemiology Study, the absolute risk of a subsequent fracture following an initial osteoporotic fracture was similar for women and men at every age group and approximately equal to that of an initial fracture risk for a woman 10-year older or a man 20-year older.[32] For example, a woman or man aged 60–69 years had a risk of a subsequent fracture that was similar to an initial fracture risk of a woman aged 70–79 or of a man aged 80+ years. Thus, overall, the relative risk of a subsequent fracture was twofold higher for women and more than threefold higher for men. Similar results have been reported from other cohorts.[30,31] In the Tromsø study, the relative risk of subsequent fracture following any type of initial fracture was higher in men (RR: 2.0, 95% CI, 1.6–2.4) than women (RR, 1.3, 95% CI, 1.2–1.5) that also implies similar absolute refracture rates between the sexes.[30] Similar subsequent fracture incidence rates between women and men have been reported in the nationwide Danish registry study for all types of initial and subsequent fractures.[31]

The Role of Type of Initial Fracture in Subsequent Fracture Risk

Although the risk of subsequent fracture is elevated following all types of initial fracture, the magnitude of this association depends on the initial fracture type. In a metaanalysis the risk of subsequent fracture was on average twofold higher than the prior osteoporotic fracture[27] with the highest magnitude of association of ~fourfold being observed for the association between prior and subsequent clinical vertebral fractures.

Initial Hip Fracture

The increased risk of subsequent fracture following an initial hip fracture is well known. The magnitude of this association depends on the study population and design, but it is essentially more than twofold compared to initial fracture risk.[27] A similar magnitude of a subsequent fracture risk following an initial hip fracture was reported in the large UK General Practice Research Database.[29] A gender-specific analysis in this study revealed a higher risk in men than women. Thus, a man aged 60–74 with an initial hip fracture had an eightfold increased risk of another fracture, whereas the corresponding risk for women was fourfold. This gender difference attenuated with age and was similar for those aged 85 or over.

The risk of a second hip fracture following an initial hip fracture has been addressed separately in several population-based studies.[31,38–40] In these studies the incidence of a second hip fracture ranged from 2.3% to 5.1% in the year following the first hip fracture with the highest rate observed in a study from Finland.[40] There was no gender difference in the risk of second hip fracture between women and men with the exception of the Framingham study[38] where the risk in men was twice as high as the risk in women. This difference, however, disappeared after adjusting for multiple covariates, indicating perhaps a difference in health status between women and men.

Initial Vertebral Fracture

It is now clear that clinical vertebral fractures are associated with significant risk of subsequent fractures.[27,29,34,41,42] The magnitude of association ranges between two- and fourfold with the strongest association being for subsequent clinical vertebral fracture.[27] It has been estimated that ~20% of women with vertebral fractures will experience a new vertebral fracture in the subsequent year.[43] The risk of a nonspine fracture following a clinical vertebral fracture is lower ranging from 1.8- to 2.7-fold depending on the characteristics of the population studied such as age and gender.[27,29] The association between clinical vertebral fracture and subsequent wrist and hip fracture has been found particularly in women.[29,44]

The association between asymptomatic vertebral fractures and new clinical vertebral fracture is also acknowledged in several studies.[45–47] The risk of sustaining a nonvertebral fracture following vertebral deformities or asymptomatic vertebral fractures ranged between 2- and 2.8-fold depending on the number and severity of vertebral deformities.[46,48]

The risk of a subsequent hip or other limb fractures following asymptomatic vertebral fractures has also been reported in several studies, particularly in women.[28,45,48–50] The characteristics of vertebral deformities that best predict subsequent fracture risk are the number of vertebral deformities, with the risk increasing from 3.2 (2.1–4.8) for one vertebral deformity to 9.8 (6.1–15.8) for two and 23.3 (15.3–35.4) for three or more vertebral deformities.[51] The severity of vertebral fracture such as a reduction in any of the vertebral body ratios by over 40% also played a role in subsequent fracture risk prediction.[52] However, in a more recent study of 3358 postmenopausal women, the presence of a prevalent mild vertebral deformity was associated with 1.8-fold of a vertebral subsequent fracture risk and 1.32-fold risk of a nonvertebral fracture.[53]

Initial Nonhip Nonvertebral Fracture

Despite the fact that NHNVs represent at least half of all the fractures occurring in the community, there is a paucity of data on the risk of refracture following these types of fractures. In the Dubbo Osteoporosis Epidemiology Study, NHNV fractures not only represented half of all low-trauma fractures that occurred during the 20 years of follow-up but also preceded half of all subsequent fractures. When considered as a group, initial NHNV fractures were associated with a 2.0–2.5-fold increased risk of any subsequent fracture, 2-fold risk of a subsequent vertebral fracture in women and men and 2-fold risk of subsequent hip fracture, particularly in women.[6]

Fractures of the upper and lower limb, and in particular distal forearm fractures, have been reported in several studies to increase the risk of any subsequent fracture[30–32,54–56] as well as hip and spine subsequent fractures.[31,57] Notably, in a metaanalysis of nine studies conducted in Europe and United States, Colles fractures were associated with a significant increased risk of subsequent hip fractures, with men having a significantly higher relative risk than women.[58] Lower limb fractures have also been associated with increased risk of any subsequent fractures or hip and vertebral subsequent fractures in several studies.[30–32] Ankle fractures have been associated with increased subsequent fracture risk in men, but not in women.[30,32] Ankle fractures, by contrast with other typical osteoporotic fracture, may be associated with other risk factors than low BMD. There is evidence, that ankle fracture, often occur as the result of a twisting force, and heavier women are at risk.[59] By contrast, rib fractures have been found to be associated with an increased subsequent fracture risk in women, but not men.[32,57]

Timing of Subsequent Fracture

The risk of a subsequent fracture is not uniform over time, with a history of a more recent prior fracture carrying substantially greater risk for subsequent fractures than a prior fracture that occurred in the more distant past.[32,43,60,61] In the Dubbo cohort, approximately 41% of fractures in women and 52% fractures in men occurred within the first 2 years following the initial fracture.[32] Based on life tables analysis, the excess refracture risk was elevated for the first 5-year post initial fracture, and then decreased progressively over the following years, becoming no longer significant beyond 10 years following the initial osteoporotic fracture.[32] These findings were similar in other studies. In a study from the Netherlands the relative risk of a subsequent fracture was 5.3, 2.8 and 1.4 at 1, 2–5 and 6–10 years postfracture, with the risk no longer significant after 10 years postfracture.[62]

The reason for this increased risk of subsequent fracture immediately following an initial osteoporotic fracture is not fully understood. Immobilisation and the associated muscle loss, risk of falling and deterioration of health may play a role.[63,64] The pattern of subsequent fracture risk with time highlights the importance of treatment initiation as soon as possible following the first osteoporotic fracture.

The pattern of subsequent fracture risk over time also varies according to the type of initial fracture. The current evidence suggests that although all types of osteoporotic fractures, including minor (or peripheral) fractures, are associated with increased subsequent fracture risk in the first 5 years, only major ones such as hip and spine fractures are associated with long-term risk of subsequent fracture up to 10 or 15 years postfracture.[65–67] In the large Danish hip fracture study, the greatest risk of a subsequent hip fracture was observed in the first year postfracture (RR 2.2; 95% CI, 2.0–2.5) and then gradually declined but did not return to the populations' initial risk until 15 years. However, by 10 years the increased risk was minor (RR, 1.01; 95% CI, 1.00–1.02).[67] In the Manitoba DXA health-care database, the risk of subsequent fracture following major fractures (i.e., hip, spine, forearm, humerus) was elevated up to 10 years postfracture, whereas for minor fractures (defined as rib, sternum, trunk, clavicle, scapula, patella, tibia/fibula, ankle) the risk of future fractures was increased only in the first year postfracture.[65]

The estimation of long-term risk of subsequent fracture in the elderly population is affected by the competing risk of mortality. Several studies have highlighted age as a modifier of time to subsequent fracture with the long-term relative risk of it being higher in younger versus older age groups as the latter groups have a higher mortality risk.[60,66,68] In the Reykjavik Study of major osteoporotic fractures, the immediate risk of subsequent fracture was similar across all age groups, but in the subsequent years, the risk remained elevated only for those aged 60 at baseline, whereas for those aged 80–90 at baseline the risk of new fractures declined markedly with time.[66] Thus, the main cause for the interaction between age and time to subsequent fracture resides mostly on the fact that older individuals have a higher mortality risk and thus do not survive to experience a subsequent fracture. The issue of the competing risk of subsequent fracture and mortality was illustrated in a large retrospective study of subsequent fracture risk and mortality in men with hip fracture over 22 years of follow-up. In this study there was a sharp difference in the absolute risk of subsequent fracture between the unadjusted and mortality-adjusted risk of subsequent fracture. This discrepancy, as expected, was even more pronounced in older individuals and increased with increasing time since the initial fracture. Thus, the majority of new subsequent fractures occurred in men younger than 75 years of age, whereas the majority of those 75 years or older died before they had a chance to experience new fracture.[68] The main message, however, is that the estimates of subsequent fracture risk, particularly for groups with greater mortality risk depend vastly on the methods employed by authors, and thus could be unreliable. On the other hand, it is important to note that osteoporotic fractures are themselves associated with increased mortality, and thus at least part of the mortality in these groups may not be really 'competing' with, but part of the outcome following an initial osteoporotic fracture.[69,70] This will be discussed further in the following section on postfracture mortality.

Role of BMD in Predicting Subsequent Fracture Risk

Although low BMD is the best predictor of first fracture risk in both women and men, its role in the risk of subsequent fracture is less clear.[71] This uncertainty comes from the fact that a large proportion of fragility fractures occur in women and men with BMD, T-score above −2.5 SD.[72–77] The proportion of fractures occurring in individuals with normal or osteopenic BMD is reportedly higher in men (up to 74%) than women (55%).[75,78,79] The risk factors contributing to fracture risk in this population have been studied in several cohorts. Decreasing BMD (from −1 SD to −2 SD), prior fracture and worse general health were independent risk predictors in the Canadian Multicentre Osteoporosis Study. In the Dubbo Osteoporosis Epidemiology Study, exposure to at least one of the known independent fracture risk factor (e.g. lower BMD, age, falls in the previous year, prior fracture) accounted for 49% of non-osteoporotic fractures in women and 39% in men.[78] In a prospective study of the participants enrolled in the Concord Secondary Fracture Programme, the risk of refracture was predicted by comorbidities, corticosteroid therapy, low femoral neck BMD and medication compliance in the group classified as high risk, whereas the presence of high levels of urinary deoxypyridinoline was the only refracture predictor in the group deemed as having low fracture risk.[80]

Several studies have attempted to quantify the risk of subsequent fracture in participants with BMD above a T-score of −2.5 SD. In the Canadian Multicentre Osteoporosis Study, women and men with osteopenia (54% women and 58% men) contributed to over 40% of all subsequent fractures (47% women and 42% men).[73] In the Dubbo Osteoporosis Epidemiology Study, the relative risk of subsequent fracture was double the risk of first fracture for each level of BMD (i.e., osteoporotic, osteopenia and normal femoral neck BMD), with the likelihood of falling and decreased quadriceps strength contributing to subsequent fracture risk in those with normal BMD.[72]

However, despite this evidence of increased fracture risk, there is still no clear clinical treatment guideline for individuals with BMD above −2.5 SD who experience fractures. Most randomised controlled trials (RCTs) of osteoporosis treatment include only participants with a T-score of −2.5 or less, or at least those with more severe osteopenia, and thus evidence of treatment efficacy in this group is lacking. However, in the FIT trial which included participants with higher BMD, alendronate had no antifracture benefit in those with a T-score of −2.0 or higher.[81]

Secular Trends for Subsequent Fracture Risk

A decline in the trend of hip fracture incidence has been recently reported for Western countries in particular.[82–84] By contrast, in most but not all[85] Asian countries such as China,[86] Japan[21] and South Korea,[87] there has been a continuous increase in hip fracture incidence. The reason behind the decline in fracture incidence in Western countries is not fully understood, but it is believed to be related on one hand to changes in life style habits which lead to obesity and consequently increase in bone mineral density and on the other hand to improvements in health services and better osteoporosis management. However, it is less clear whether the decline in fracture incidence is also associated with reduced risk of subsequent fracture.

The secular trend of subsequent fracture following hip fracture was reported using electronic health care databases in Norway, United Kingdom, and United States.[24,88,89] The Norwegian and UK study reported stable or increasing rates of recurrent hip or major nonhip fracture.[88,90] By contrast in a small population-based study conducted in United States between 1980 and 2006, the decline in hip fracture recurrence was even greater than the decline in the initial hip fracture rate.[89]

An Australian study investigated the secular trend of subsequent fracture rates following all types of osteoporotic fractures using two birth cohorts over two nonoverlapping decades, 1989/1999 and 2000/2010, in an effort to capture both environmental and cohort specific factors.[91] Despite a decline in the secular trend of initial osteoporotic fracture over those decades, the absolute rate of subsequent fracture remained stable over the two decades for both women and men. Although the decline of the initial fracture rate seemed to be related to a greater proportion of obesity and hence greater BMD in the second cohort, there was no apparent reason for the persistently high absolute refracture rate.

In summary, following any initial fracture, there is an overall twofold increased risk of a subsequent fracture with the greatest risk occurring close in time to the initial fracture. The risk is higher in men than women, occurs for axial and peripheral fractures and, despite the declining incidence of at least initial hip fracture, does not appear to have changed over time.

RISK OF MORTALITY FOLLOWING OSTEOPOROTIC FRACTURES

Premature mortality following hip, vertebral and other major fractures (e.g., humerus, distal femur, and pelvis) is now well recognised. There is also emerging evidence that even minor fractures (e.g., forearm) are associated with increased mortality risk in the elderly.[69,92–94] However, the increased postfracture mortality risk is not constant over time. Evidence from numerous epidemiological studies suggests that excess mortality is highest close to the fracture event. Except for hip and perhaps vertebral fractures, for which excess mortality persists for at least 10 years, there is no excess mortality beyond 5 years for the other fracture types.[58,95,96]

Overall, the magnitude of the fracture mortality association ranges between 15% to over twofold above the age- and gender-expected population mortality risk, depending mostly on the timing from and type of initial osteoporotic fracture.[69,93,97,98]

The mechanism for mortality risk is most likely multifactorial. The number and severity of comorbidities at the time of fracture may play a role in postfracture mortality; however, the evidence is not consistent amongst studies.[99,100] Several risk factors for fracture such as low bone mass, bone loss,[101,102] muscle weakness[103,104] and sarcopenia[105] have also been found to be associated with mortality risk independent of fracture and thus could also contribute to postfracture mortality risk.

Mortality Risk According to Fracture Type

Hip Fracture

Hip fractures are associated with the highest mortality rates amongst all osteoporotic fractures.[69,97,98] The magnitude of hip fracture mortality depends on the time from the fracture and age. Up to 30% of individuals with hip fracture are estimated to die within the first 2 years postfracture.[106,107] This equates to an excess mortality of over twofold the expected general population mortality.[58,108] The yearly excess mortality from 22 cohorts in women and 17 in men was recently estimated in a metaanalysis.[58] Using a life table approach, investigators estimated an annual posthip fracture excess mortality of 8%, 11%, 18% and 22% at 1, 2, 5 and 10 years for women with hip fracture at age 80 compared to women without fracture. In men, the estimated annual posthip fracture excess mortality was even higher: 18%, 22%, 26% and 20% for 80-year-old men with hip fracture compared with those without fracture.

Although hip fracture occurs predominantly in the older age groups, excess mortality is higher in younger individuals when compared with their nonfractured peers than older individuals. For example, in the Dubbo Osteoporosis Epidemiology Study, younger women with hip fracture had a >eightfold increased mortality risk compared with a twofold increased risk for those older than 75 years.[69] Similar data on higher excess mortality in younger women with hip fracture versus older women have also been reported in other studies, even after adjusting for comorbidities and other risk factors.

Vertebral Fracture

Vertebral fractures are the most common type of osteoporotic fracture, representing a third of all osteoporotic fractures,[4] and are associated with significant morbidity and mortality.[109,110] However, despite their severity there is far less data on the mortality associated with vertebral fractures compared with that of hip fracture. Nevertheless, the majority of,[69,92,97,111] although not all,[93] studies that have investigated outcomes following vertebral fractures reported increased mortality risk.

Research on vertebral fractures poses a significant challenge because of the fact that only about 30% of them are symptomatic or admitted to hospital.[10,112] Symptomatic vertebral fractures are associated with increased risk of mortality in both women and men, with a magnitude ranging from 32% to over 13-fold above an expected general population mortality.[97,109,110]

Similar to hip fracture mortality, age plays a role in the relative fracture mortality magnitude. In the Dubbo Osteoporosis Epidemiology Study, individuals who experienced a symptomatic vertebral fracture below 75 years of age had significantly higher excess mortality compared with those who had their first vertebral fracture after 75 years for both women (age-adjusted SMR 3.8 (2.5–5.8) and 1.5 (1.1–1.8)) and men (age-adjusted SMR 4.2 (2.4–7.3) and 1.9 (1.4–2.6)).[69]

The effect of asymptomatic vertebral fractures on mortality risk is less well studied. In the initial and follow-up studies from the Dubbo Osteoporosis Epidemiology Study, asymptomatic vertebral fractures were associated with a similar increased mortality risk as symptomatic fractures.[69,98] Certainly the presence of multiple vertebral deformities has been found to be associated with increased mortality risk in several studies.[48,113,114] In another study from Dubbo Osteoporosis Epidemiology Study, vertebral deformities were associated with high risk of incident symptomatic vertebral fracture which increased mortality risk.[46] The role of a single vertebral deformity on mortality is less clear with contradictory findings,[114,115] and any effect may be mediated through increasing risk of new clinical vertebral fractures.

Nonhip Nonvertebral Fracture

NHNV fractures are the most common group of osteoporotic fractures.[6] However, despite their frequency there are very few studies on the adverse outcomes associated with NHNV fractures. Increased mortality risk associated with major NHNV fractures (pelvis, distal femur, multiple ribs and humerus) was first demonstrated in the Dubbo Osteoporosis Epidemiology Study over a 5-year follow-up.[98] Major NHNV fractures were associated with 2.0–2.2-fold increased mortality risk for both women and men, whereas minor NHNV fractures (forearm, wrist, metacarpal, ankle and metatarsal) were not associated with increased mortality risk. Subsequently, a long-term follow-up study (>18 years) from the same cohort reconfirmed the excess mortality associated with major NHNV fracture and also demonstrated increased mortality for minor NHNV fractures in individuals older than 75 years (37–82% increase in mortality risk).[69] Notably, given that NHNV fractures represented half of all incident fractures in the cohort, they were responsible for over 40% of all deaths and contributed to a third of all the excess mortality. Similarly, high mortality risk associated with major NHNV fracture was reported in a hospital database study from Maastricht.[116] In the Canadian Multicentre Osteoporosis Study, all NHNV fractures were associated with 38% significant mortality risk increase after adjusting for all risk factors in women, but not in men.[117] Of the NHNV fractures, ribs, humerus, forearm and pelvis were associated with increased mortality risk in women, whereas in men only humerus fractures were associated with it.[117] Proximal humerus, rib, pelvis and distal femur fractures have also been reported by others to be associated with increased mortality risk.[92–94,118,119] Mortality after distal forearm fractures, on the other hand, is controversial with some[118,120]; but not all[92–94] studies reporting increased risk.

The Role of Time Post Initial Fracture

Several long-term epidemiological prospective studies have reported the pattern of postfracture mortality risk over time.[69,92,116,118] Postfracture mortality is highest in the first 5 years postfracture and then declines towards the general population mortality rates except for hip fracture where the increased mortality lasts for at least 10 years.[69,92,116,118]

The pattern of mortality with time has been best described for hip fractures. Data on excess posthip fracture mortality over time were examined in a recent metaanalysis. In all included studies, the magnitude of mortality risk was very high immediately postfracture and then decreased afterwards, but it did not reach the expected background mortality for up to 10 years postfracture for any given age. All-cause excess mortality ranged between 5- and 8-fold at 3 months, then declined to 2.9–3.7-fold at 1 year and to 1.8–2.0-fold at 10 years postfracture.[58] A similar pattern of excess mortality over the time was found in a metaanalysis of 22 studies. In this review mortality rates ranged between 8.4% and 36% in the first year postfracture, which corresponded to up to twofold above the expected population mortality.[108]

A similar pattern of high excess mortality immediate after the fracture has also been described for vertebral fracture in several studies.[69,92,115,118] The pattern of mortality risk following symptomatic vertebral fractures is similar to that following hip fracture with the highest mortality recorded in the first 1–5 years postfracture.[69,121]

For other fracture types, there is far less data. In the Dubbo Osteoporosis Epidemiology Study, the risk of mortality was elevated for the first 5 years postfracture for all types of fracture, including minor fractures in those older than 75 years.[69] Beyond 5 years, mortality risk was elevated only for hip fracture, and beyond 10 years none of the fractures were associated with excess mortality.

Subsequent Fracture and Mortality Risk

The role of subsequent fracture on mortality risk was first evaluated in the Dubbo Osteoporosis Epidemiology Study by comparing the pattern of excess mortality over time for those with one versus more than one fracture.[69] A subsequent fracture increased that initial 5-year mortality risk by up to twofold for both women and men. The excess mortality following the second fracture was higher than the excess mortality solely attributed to the initial fracture resulting in an excess mortality persisting for up to 10 years in both women and men.[69] Interestingly, a subsequent study from the same cohort demonstrated that the majority of the excess mortality beyond 5 years postfracture was attributable to the subsequent fracture, whereas those who survived for 10 years without experiencing a second fracture were at low risk of experiencing either another fracture or premature mortality.[70] The role of subsequent fracture in mortality risk has also been reported in a Dutch study where a subsequent fracture was an independent predictor of mortality risk after adjusting for age, gender and fracture type.[116]

Understanding of the role of subsequent fracture for mortality risk is particularly important given its high frequency following all types of fracture and its potential to be prevented by antifracture therapy.

Secular Trends in Postfracture Mortality Risk

The last two decades have been characterised by a significant increase in life expectancy, perhaps because of better management of chronic diseases and cancer. However, there is still ample evidence that the management of fracture in the community is poor (<30% in women and <20% in men). Recent research suggested that although absolute mortality following hip fracture has plateaued or even declined[122–125], with the increasing life expectancy in the population, the gap between hip fracture and general population mortality has not decreased over time.[91,122,126,127] Similarly, although a reduction in absolute postfracture mortality was reported for all types of osteoporotic fractures over a 20-year period, the relative mortality of the fracture population compared with the general population mortality had not declined.[91]

The Association Between Antiresorptive Medication and Mortality Risk Reduction

Bisphosphonates are currently considered as first-line treatment for osteoporosis worldwide.[128,129] There is abundant evidence that these agents effectively reduce the risk of both vertebral and nonvertebral fractures.[130–132] More recently, bisphosphonates have also been linked to improved survival in both RCTs[133,134] as well as observational studies.[135–139]

Evidence From Randomised Controlled Trials

The first evidence of a positive association between bisphosphonate use and survival came from an RCT of zoledronic acid in patients with hip fracture.[134] In this trial, zoledronic acid, as well as resulting in a significant reduction in the rate of both vertebral and nonvertebral fracture, also resulted in a 28% mortality risk reduction compared to placebo. Subsequently, all placebo-controlled RCTs of approved osteoporosis medications, including eight trials of risedronate, zoledronic acid, denosumab and strontium ranelate were included into a metaanalysis.[133] The pooled survival benefit in this metaanalysis was ~11% (RR, 0.89 (95% CI, 0.80–0.99)). A higher baseline mortality rate in the control group was the only factor associated with a greater survival benefit, whereas age, fracture type or magnitude of fracture risk reduction did not contribute significantly. Notably, in this metaanalysis, although the association between osteoporosis medication and survival suggested survival benefit, the only RCT that individually demonstrated a significant association with survival was the zoledronic acid study described above. It is important to note that the zoledronic acid study had much broader entry criteria than the other RCTs and thus had a significantly higher mortality rate in the placebo arm. In the other studies with lower placebo mortality, there was limited power to detect any relationship between mortality and treatment.

Evidence From Observational Studies

A significant positive association between bisphosphonates and mortality risk has also been reported in several observational osteoporosis cohorts[137–140] and registry-based studies.[136,141] Although the majority of these studies are restricted to hip fracture,[136,138,141,142] there is also some evidence of survival benefit for vertebral fracture[143] and all types of fracture.[137] The magnitude of the fracture mortality association varies widely amongst these studies and ranged between 24% and 69%, perhaps because of differences in study populations and length of follow-up.

Studies that have included both women and men suggest that the treatment-related survival benefit may be stronger in women than men.[136,137,141] In the Danish National registry study there was an interaction between treatment use and gender, with mortality risk being reduced in women (adjusted OR, 071; 95% CI, 053–0.94) but not in men (OR, 0.82; 95% CI, 0.41–1.64).[136] In the Dubbo Osteoporosis Epidemiology Study, the positive association between treatment and survival was observed for both genders, but it was statistically significant only in women (adjusted HR, 0.31; 95% CI, 0.17–0.59) and not in men (adjusted HR, 0.48; 95% CI, 0.11–1.98). The reason for the weaker effect in men is not clear, but it is possible that the lower number of medication users may play a role. The gender discrepancy in treatment uptake following a hip fracture was recently reported in an Austrian Nationwide Study.[141] Initiation of bisphosphonates was associated with significant mortality risk reduction in women at all time points postfracture (90 days, 1 and 3 years) and for all age groups (except the 50–59-year age group), whereas for men mortality risk reduction was statistically significant when all age groups were analysed together, but not in individual age-group strata, possibly related to the small number of men on treatment.

The mechanism through which bisphosphonates may reduce mortality risk following fracture is not entirely clear and is likely to be multifactorial. The most obvious mechanism would be through a reduction in subsequent fracture risk. However, a reduction in risk of subsequent fracture contributed only modestly, ~8%, to mortality risk reduction in the RCT of zoledronic acid,[144] but did not mediate mortality reduction in any of the observational study.[137,138,141]

It is also possible that the relationship between bisphosphonates and mortality risk may be mediated through a reduction in the rate of bone loss. Accelerated bone loss has been reported to be associated with an increase in mortality risk in both individuals with[101] and without fracture.[102,145] High bone loss has previously been reported to be a marker of poor health and increased mortality.[146] One theory as to why a decrease in bone loss may have an effect on mortality risk is through a reduction in the release of toxic substances, such as lead, that are normally sequestered in bone. Several other theories have been postulated including effects on inflammation and immune function.[147,148] However, at the present time there is no direct evidence for any of these theories.

CONCLUSION

Osteoporotic fractures remain a significant and growing burden for both the society (through direct and indirect costs) and individuals through the associated risk of further fracture, disability and premature mortality. The segment of population at fracture risk is growing rapidly, and even if the decline observed in the rate of hip fracture is confirmed for other fracture types, the total burden of fracture is still expected to increase. Most importantly, the severe outcomes associated with fracture have not followed the same downward trend.

Although now well documented, the outcomes of subsequent fracture and premature mortality are not well appreciated. Both of these risks are highest immediately after the initial fracture. The risk of subsequent fracture is increased for virtually all osteoporotic fractures and for all levels of BMD. Mortality risk is increased following all major fracture such as hip, clinical vertebral and other proximal fractures, and there is growing evidence that minor fracture such as distal forearm fracture are also associated with increased mortality risk in older individuals. A subsequent fracture is associated with a further excess mortality risk, persisting for at least 5 years beyond that subsequent fracture. Importantly, there is a growing body of evidence from both an RCT and cohort studies that antiresorptive medication may be associated with improved survival in patients with increased fracture risk.

Thus, early evaluation and treatment of all fragility fracture is vital to at least reduce the burden of subsequent fractures and possibly decrease the associated mortality risk.

REFERENCES

[1] NIH. Osteoporosis prevention, diagnosis, and therapy. *NIH Consens Statement* 2000;**17**:1–45.

[2] Johnell O, Kanis JA. An estimate of the worldwide prevalence and disability associated with osteoporotic fractures. *Osteoporos Int* 2006;**17**:1726–33.

[3] Hernlund E, Svedbom A, Ivergard M, Compston J, Cooper C, Stenmark J, Mccloskey EV, Jonsson B, Kanis JA. Osteoporosis in the European Union: medical management, epidemiology and economic burden. A report prepared in collaboration with the International Osteoporosis Foundation (IOF) and the European Federation of Pharmaceutical Industry Associations (EFPIA). *Arch Osteoporos* 2013;**8**:136.

[4] Burge R, Dawson-Hughes B, Solomon DH, Wong JB, King A, Tosteson A. Incidence and economic burden of osteoporosis-related fractures in the United States, 2005-2025. *J Bone Miner Res* 2007;**22**:465–75.

[5] Osteoporosis-Australia. *The burden of brittle bones epidemiology, costs & burden of osteoporosis in Australia -2007*. 2007.

[6] Bliuc D, Nguyen TV, Eisman JA, Center JR. The impact of nonhip nonvertebral fractures in elderly women and men. *J Clin Endocrinol Metab* 2014;**99**:415–23.

[7] Johnell O, Kanis J. Epidemiology of osteoporotic fractures. *Osteoporos Int* 2005;**16**(Suppl. 2):S3–7.

[8] Chang KP, Center JR, Nguyen TV, Eisman JA. Incidence of hip and other osteoporotic fractures in elderly men and women: Dubbo Osteoporosis Epidemiology Study. *J Bone Miner Res* 2004;**19**:532–6.

[9] Kanis JA, Odén A, Mccloskey EV, Johansson H, Wahl DA, Cooper C. A systematic review of hip fracture incidence and probability of fracture worldwide. *Osteoporos Int* 2012;**23**:2239–56.

[10] Cooper C, Atkinson EJ, O'fallon WM, Melton 3rd LJ. Incidence of clinically diagnosed vertebral fractures: a population-based study in Rochester, Minnesota, 1985-1989. *J Bone Miner Res* 1992;**7**:221–7.

[11] Delmas PD, Van De Langerijt L, Watts NB, Eastell R, Genant H, Grauer A, Cahall DL. Underdiagnosis of vertebral fractures is a worldwide problem: the IMPACT study. *J Bone Miner Res* 2005;**20**:557–63.

[12] Grados F, Marcelli C, Dargent-Molina P, Roux C, Vergnol JF, Meunier PJ, Fardellone P. Prevalence of vertebral fractures in French women older than 75 years from the EPIDOS study. *Bone* 2004;**34**:362–7.

[13] Jackson SA, Tenenhouse A, Robertson L. Vertebral fracture definition from population-based data: preliminary results from the Canadian Multicenter Osteoporosis Study (CaMos). *Osteoporos Int* 2000;**11**:680–7.

[14] Felsenberg D, Silman AJ, Lunt M, Armbrecht G, Ismail AA, Finn JD, Cockerill WC, Banzer D, Benevolenskaya LI, Bhalla A, Bruges Armas J, Cannata JB, Cooper C, Dequeker J, Eastell R, Felsch B, Gowin W, Havelka S, Hoszowski K, Jajic I, Janott J, Johnell O, Kanis JA, Kragl G, Lopes Vaz A, Lorenc R, Lyritis G, Masaryk P, Matthis C, Miazgowski T, Parisi G, Pols HA, Poor G, Raspe HH, Reid DM, Reisinger W, Schedit-Nave C, Stepan JJ, Todd CJ, Weber K, Woolf AD, Yershova OB, Reeve J, O'neill TW. Incidence of vertebral fracture in Europe: results from the European Prospective Osteoporosis Study (EPOS). *J Bone Miner Res* 2002;**17**:716–24.

[15] Van der Klift M, De Laet CE, Mccloskey EV, Hofman A, Pols HA. The incidence of vertebral fractures in men and women: the Rotterdam Study. *J Bone Miner Res* 2002;**17**:1051–6.

[16] Cauley JA, Hochberg MC, Lui LY, Palermo L, Ensrud KE, Hillier TA, Nevitt MC, Cummings SR. Long-term risk of incident vertebral fractures. *JAMA* 2007;**298**:2761–7.

[17] O'Neill TW, Felsenberg D, Varlow J, Cooper C, Kanis JA, Silman AJ. The prevalence of vertebral deformity in european men and women: the European Vertebral Osteoporosis Study. *J Bone Miner Res* 1996;**11**:1010–8.

[18] Kanis JA, McCloskey EV. Epidemiology of vertebral osteoporosis. *Bone* 1992;**13**(Suppl. 2):S1–10.

[19] Adams AL, Shi J, Takayanagi M, Dell RM, Funahashi TT, Jacobsen SJ. Ten-year hip fracture incidence rate trends in a large California population, 1997-2006. *Osteoporos Int* 2013;**24**:373–6.

[20] Cooper C, Cole ZA, Holroyd CR, Earl SC, Harvey NC, Dennison EM, Melton LJ, Cummings SR, Kanis JA, The IOF CSA Working Group on Fracture Epidemiology, Adachi J, Borgstrom F, Dimai HP, Clark P, Lau E, Lewiecki EM, Lips P, Lorenc R, Mccloskey E, Ortolani S, Papaioannou A, Silverman S, Wahl DA, Yoshimura N. Secular trends in the incidence of hip and other osteoporotic fractures. *Osteoporos Int* 2011;**22**:1277–88.

[21] Hagino H, Furukawa K, Fujiwara S, Okano T, Katagiri H, Yamamoto K, Teshima R. Recent trends in the incidence and lifetime risk of hip fracture in Tottori, Japan. *Osteoporos Int* 2009;**20**:543–8.

[22] Lau EM, Cooper C, Fung H, Lam D, Tsang KK. Hip fracture in Hong Kong over the last decade–a comparison with the UK. *J Public Health Med* 1999;**21**:249–50.

[23] Gullberg B, Johnell O, Kanis JA. World-wide projections for hip fracture. *Osteoporos Int* 1997;**7**:407–13.

[24] Omsland TK, Holvik K, Meyer HE, Center JR, Emaus N, Tell GS, Schei B, Tverdal A, Gjesdal CG, Grimnes G, Forsmo S, Eisman JA, Sogaard AJ. Hip fractures in Norway 1999-2008: time trends in total incidence and second hip fracture rates: a NOREPOS study. *Eur J Epidemiol* 2012;**27**:807–14.

[25] Hodsman AB, Leslie WD, Tsang JF, Gamble GD. 10-year probability of recurrent fractures following wrist and other osteoporotic fractures in a large clinical cohort: an analysis from the Manitoba Bone Density Program. *Arch Intern Med* 2008;**168**:2261–7.

[26] Kanis JA, Johansson H, Oden A, Johnell O, De Laet C, Eisman JA, Mccloskey EV, Mellstrom D, Melton 3rd LJ, Pols HA, Reeve J, Silman AJ, Tenenhouse A. A family history of fracture and fracture risk: a meta-analysis. *Bone* 2004a;**35**:1029–37.

[27] Klotzbuecher CM, Ross PD, Landsman PB, Abbott 3rd TA, Berger M. Patients with prior fractures have an increased risk of future fractures: a summary of the literature and statistical synthesis. *J Bone Miner Res* 2000;**15**:721–39.

[28] Schousboe JT, Fink HA, Lui LY, Taylor BC, Ensrud KE. Association between prior non-spine non-hip fractures or prevalent radiographic vertebral deformities known to be at least 10 years old and incident hip fracture. *J Bone Miner Res* 2006;**21**:1557–64.

[29] van Staa TP, Leufkens HG, Cooper C. Does a fracture at one site predict later fractures at other sites? A British cohort study. *Osteoporos Int* 2002;**13**:624–9.

[30] Ahmed LA, Center JR, Bjornerem A, Bluic D, Joakimsen RM, Jorgensen L, Meyer HE, Nguyen ND, Nguyen TV, Omsland TK, Stormer J, Tell GS, Van Geel TA, Eisman JA, Emaus N. Progressively increasing fracture risk with advancing age after initial incident fragility fracture: the Tromso study. *J Bone Miner Res* 2013;**28**(10):2214–21.

[31] Hansen L, Petersen KD, Eriksen SA, Langdahl BL, Eiken PA, Brixen K, Abrahamsen B, Jensen JE, Harslof T, Vestergaard P. Subsequent fracture rates in a nationwide population-based cohort study with a 10-year perspective. *Osteoporos Int* 2015;**26**:513–9.

[32] Center JR, Bliuc D, Nguyen TV, Eisman JA. Risk of subsequent fracture after low-trauma fracture in men and women. *JAMA* 2007;**297**:387–94.

[33] Cummings SR, Nevitt MC, Browner WS, Stone K, Fox KM, Ensrud KE, Cauley J, Black D, Vogt TM. Risk factors for hip fracture in white women. Study of Osteoporotic Fractures Research Group. *N Engl J Med* 1995;**332**:767–73.

[34] Lauritzen JB, Lund B. Risk of hip fracture after osteoporosis fractures. 451 women with fracture of lumbar spine, olecranon, knee or ankle. *Acta Orthop Scand* 1993;**64**:297–300.

[35] Mussolino ME, Looker AC, Madans JH, Langlois JA, Orwoll ES. Risk factors for hip fracture in white men: the NHANES I Epidemiologic Follow-up Study. *J Bone Miner Res* 1998;**13**:918–24.

[36] Seeley DG, Kelsey J, Jergas M, Nevitt MC. Predictors of ankle and foot fractures in older women. The Study of Osteoporotic Fractures Research Group. *J Bone Miner Res* 1996;**11**:1347–55.

[37] Doherty DA, Sanders KM, Kotowicz MA, Prince RL. Lifetime and five-year age-specific risks of first and subsequent osteoporotic fractures in postmenopausal women. *Osteoporos Int* 2001;**12**:16–23.

[38] Berry SD, Samelson EJ, Hannan MT, Mclean RR, Lu M, Cupples LA, Shaffer ML, Beiser AL, Kelly-Hayes M, Kiel DP. Second hip fracture in older men and women: the Framingham Study. *Arch Intern Med* 2007;**167**:1971–6.

[39] Chapurlat RD, Bauer DC, Nevitt M, Stone K, Cummings SR. Incidence and risk factors for a second hip fracture in elderly women. The Study of Osteoporotic Fractures. *Osteoporos Int* 2003;**14**:130–6.

[40] Lonnroos E, Kautiainen H, Karppi P, Hartikainen S, Kiviranta I, Sulkava R. Incidence of second hip fractures. A population-based study. *Osteoporos Int* 2007;**18**:1279–85.

[41] Gunnes M, Mellstrom D, Johnell O. How well can a previous fracture indicate a new fracture? A questionnaire study of 29,802 postmenopausal women. *Acta Orthop Scand* 1998;**69**:508–12.

[42] Melton 3rd LJ, Atkinson EJ, Cooper C, O'fallon WM, Riggs BL. Vertebral fractures predict subsequent fractures. *Osteoporos Int* 1999;**10**:214–21.

[43] Lindsay R, Silverman SL, Cooper C, Hanley DA, Barton I, Broy SB, Licata A, Benhamou L, Geusens P, Flowers K, Stracke H, Seeman E. Risk of new vertebral fracture in the year following a fracture. *JAMA* 2001;**285**:320–3.

[44] Haentjens P, Johnell O, Kanis JA, Bouillon R, Cooper C, Lamraski G, Vanderschueren D, Kaufman JM, Boonen S. Evidence from data searches and life-table analyses for gender-related differences in absolute risk of hip fracture after Colles' or spine fracture: Colles' fracture as an early and sensitive marker of skeletal fragility in white men. *J Bone Miner Res* 2004;**19**:1933–44.

[45] Black DM, Arden NK, Palermo L, Pearson J, Cummings SR. Prevalent vertebral deformities predict hip fractures and new vertebral deformities but not wrist fractures. Study of Osteoporotic Fractures Research Group. *J Bone Miner Res* 1999;**14**:821–8.

[46] Pongchaiyakul C, Nguyen ND, Jones G, Center JR, Eisman JA, Nguyen TV. Asymptomatic vertebral deformity as a major risk factor for subsequent fractures and mortality: a long-term prospective study. *J Bone Miner Res* 2005;**20**:1349–55. Epub 2005 Mar 21.
[47] Ross PD, Davis JW, Epstein RS, Wasnich RD. Pre-existing fractures and bone mass predict vertebral fracture incidence in women. *Ann Intern Med* 1991;**114**:919–23.
[48] Hasserius R, Karlsson MK, Nilsson BE, Redlund-Johnell I, Johnell O. Prevalent vertebral deformities predict increased mortality and increased fracture rate in both men and women: a 10-year population-based study of 598 individuals from the Swedish cohort in the European Vertebral Osteoporosis Study. *Osteoporos Int* 2003;**14**:61–8.
[49] Fujiwara S, Kasagi F, Yamada M, Kodama K. Risk factors for hip fracture in a Japanese cohort. *J Bone Miner Res* 1997;**12**:998–1004.
[50] Ismail AA, Cockerill W, Cooper C, Finn JD, Abendroth K, Parisi G, Banzer D, Benevolenskaya LI, Bhalla AK, Armas JB, Cannata JB, Delmas PD, Dequeker J, Dilsen G, Eastell R, Ershova O, Falch JA, Felsch B, Havelka S, Hoszowski K, Jajic I, Kragl U, Johnell O, Lopez Vaz A, Lorenc R, Lyritis G, Marchand F, Masaryk P, Matthis C, Miazgowski T, Pols HA, Poor G, Rapado A, Raspe HH, Reid DM, Reisinger W, Janott J, Scheidt-Nave C, Stepan J, Todd C, Weber K, Woolf AD, Ambrecht G, Gowin W, Felsenberg D, Lunt M, Kanis JA, Reeve J, Silman AJ, O'neill TW. Prevalent vertebral deformity predicts incident hip though not distal forearm fracture: results from the European Prospective Osteoporosis Study. *Osteoporos Int* 2001;**12**:85–90.
[51] Lunt M, O'neill TW, Felsenberg D, Reeve J, Kanis JA, Cooper C, Silman AJ. Characteristics of a prevalent vertebral deformity predict subsequent vertebral fracture: results from the European Prospective Osteoporosis Study (EPOS). *Bone* 2003;**33**:505–13.
[52] Delmas PD, Genant HK, Crans GG, Stock JL, Wong M, Siris E, Adachi JD. Severity of prevalent vertebral fractures and the risk of subsequent vertebral and nonvertebral fractures: results from the MORE trial. *Bone* 2003;**33**:522–32.
[53] Roux C, Fechtenbaum J, Kolta S, Briot K, Girard M. Mild prevalent and incident vertebral fractures are risk factors for new fractures. *Osteoporos Int* 2007;**18**:1617–24.
[54] Barrett-Connor E, Sajjan SG, Siris ES, Miller PD, Chen YT, Markson LE. Wrist fracture as a predictor of future fractures in younger versus older postmenopausal women: results from the National Osteoporosis Risk Assessment (NORA). *Osteoporos Int* 2008;**19**:607–13.
[55] Honkanen R, Tuppurainen M, Kroger H, Alhava E, Puntila E. Associations of early premenopausal fractures with subsequent fractures vary by sites and mechanisms of fractures. *Calcif Tissue Int* 1997;**60**:327–31.
[56] Lauritzen JB, Schwarz P, Mcnair P, Lund B, Transbol I. Radial and humeral fractures as predictors of subsequent hip, radial or humeral fractures in women, and their seasonal variation. *Osteoporos Int* 1993;**3**:133–7.
[57] Ismail AA, Silman AJ, Reeve J, Kaptoge S, O'neill TW. Rib fractures predict incident limb fractures: results from the European prospective osteoporosis study. *Osteoporos Int* 2006;**17**:41–5.
[58] Haentjens P, Magaziner J, Colon-Emeric CS, Vanderschueren D, Milisen K, Velkeniers B, Boonen S. Meta-analysis: excess mortality after hip fracture among older women and men. *Ann Intern Med* 2010;**152**:380–90.
[59] Hasselman CT, Vogt MT, Stone KL, Cauley JA, Conti SF. Foot and ankle fractures in elderly white women. Incidence and risk factors. *J Bone Joint Surg Am* 2003;**85-A**:820–4.
[60] Johnell O, Kanis JA, Oden A, Sernbo I, Redlund-Johnell I, Petterson C, De Laet C, Jonsson B. Fracture risk following an osteoporotic fracture. *Osteoporos Int* 2004;**15**:175–9.
[61] van Helden S, Cals J, Kessels F, Brink P, Dinant GJ, Geusens P. Risk of new clinical fractures within 2 years following a fracture. *Osteoporos Int* 2006;**17**:348–54.

[62] van Geel TA, Van Helden S, Geusens PP, Winkens B, Dinant GJ. Clinical subsequent fractures cluster in time after first fractures. *Ann Rheum Dis* 2009;**68**:99–102.

[63] Bonafede M, Shi N, Barron R, Li X, Crittenden DB, Chandler D. Predicting imminent risk for fracture in patients aged 50 or older with osteoporosis using US claims data. *Arch Osteoporos* 2016;**11**:26.

[64] Van Helden S, Wyers CE, Dagnelie PC, Van Dongen MC, Willems G, Brink PR, Geusens PP. Risk of falling in patients with a recent fracture. *BMC Musculoskelet Disord* 2007;**8**:55.

[65] Giangregorio LM, Leslie WD. Time since prior fracture is a risk modifier for 10-year osteoporotic fractures. *J Bone Miner Res* 2010;**25**:1400–5.

[66] Johansson H, Siggeirsdottir K, Harvey NC, Oden A, Gudnason V, Mccloskey E, Sigurdsson G, Kanis JA. Imminent risk of fracture after fracture. *Osteoporos Int* 2017;**28**:775–80.

[67] Ryg J, Rejnmark L, Overgaard S, Brixen K, Vestergaard P. Hip fracture patients at risk of second hip fracture: a nationwide population-based cohort study of 169,145 cases during 1977–2001. *J Bone Miner Res* 2009;**24**:1299–307.

[68] von Friesendorff M, Mcguigan FE, Besjakov J, Åkesson K. Hip fracture in men—survival and subsequent fractures: a cohort study with 22-year follow-up. *J Am Geriatr Soc* 2011;**59**:806–13.

[69] Bliuc D, Nguyen ND, Milch VE, Nguyen TV, Eisman JA, Center JR. Mortality risk associated with low-trauma osteoporotic fracture and subsequent fracture in men and women. *JAMA* 2009;**301**:513–21.

[70] Bliuc D, Nguyen ND, Nguyen TV, Eisman JA, Center JR. Compound risk of high mortality following osteoporotic fracture and refracture in elderly women and men. *J Bone Miner Res* 2013;**28**:2317–24.

[71] Kanis JA, Johnell O, De Laet C, Johansson H, Oden A, Delmas P, Eisman J, Fujiwara S, Garnero P, Kroger H, Mccloskey EV, Mellstrom D, Melton LJ, Pols H, Reeve J, Silman A, Tenenhouse A. A meta-analysis of previous fracture and subsequent fracture risk. *Bone* 2004b;**35**:375–82.

[72] Bliuc D, Alarkawi D, Nguyen TV, Eisman JA, Center JR. Risk of subsequent fractures and mortality in elderly women and men with fragility fractures with and without osteoporotic bone density: the Dubbo Osteoporosis Epidemiology Study. *J Bone Miner Res* 2015a;**30**:637–46.

[73] Langsetmo L, Goltzman D, Kovacs CS, Adachi JD, Hanley DA, Kreiger N, Josse R, Papaioannou A, Olszynski WP, Jamal SA. Repeat low-trauma fractures occur frequently among men and women who have osteopenic BMD. *J Bone Miner Res* 2009;**24**:1515–22.

[74] Nguyen ND, Pongchaiyakul C, Center JR, Eisman JA, Nguyen TV. Identification of high-risk individuals for hip fracture: a 14-year prospective study. *J Bone Miner Res* 2005;**20**:1921–8.

[75] Pasco JA, Seeman E, Henry MJ, Merriman EN, Nicholson GC, Kotowicz MA. The population burden of fractures originates in women with osteopenia, not osteoporosis. *Osteoporos Int* 2006;**17**:1404–9.

[76] Schuit SC, Van Der Klift M, Weel AE, De Laet CE, Burger H, Seeman E, Hofman A, Uitterlinden AG, Van Leeuwen JP, Pols HA. Fracture incidence and association with bone mineral density in elderly men and women: the Rotterdam Study. *Bone* 2004;**34**:195–202.

[77] Siris ES, Miller PD, Barrett-Connor E, Faulkner KG, Wehren LE, Abbott TA, Berger ML, Santora AC, Sherwood LM. Identification and fracture outcomes of undiagnosed low bone mineral density in postmenopausal women: results from the National Osteoporosis Risk Assessment. *JAMA* 2001;**286**:2815–22.

[78] Nguyen ND, Eisman JA, Center JR, Nguyen TV. Risk factors for fracture in nonosteoporotic men and women. *J Clin Endocrinol Metab* 2007b;**92**:955–62.

[79] Sornay-Rendu E, Munoz F, Garnero P, Duboeuf F, Delmas PD. Identification of osteopenic women at high risk of fracture: the OFELY study. *J Bone Miner Res* 2005;**20**:1813–9.

[80] Ganda K, Schaffer A, Seibel MJ. Predictors of re-fracture amongst patients managed within a secondary fracture prevention program: a 7-year prospective study. *Osteoporos Int* 2015;**26**:543–51.

[81] Cummings SR, Black DM, Thompson DE, Applegate WB, Barrett-Connor E, Musliner TA, Palermo L, Prineas R, Rubin SM, Scott JC, Vogt T, Wallace R, Yates AJ, Lacroix AZ. Effect of alendronate on risk of fracture in women with low bone density but without vertebral fractures: results from the Fracture Intervention Trial. *JAMA* 1998;**280**:2077–82.

[82] Bergström U, Jonsson H, Gustafson Y, Pettersson U, Stenlund H, Svensson O. The hip fracture incidence curve is shifting to the right. *Acta Orthop* 2009;**80**:520–4.

[83] Chevalley T, Guilley E, Herrmann FR, Hoffmeyer P, Rapin CH, Rizzoli R. Incidence of hip fracture: reversal of a secular trend. *Rev Med Suisse* 2007;**3**. 1528–1530, 1532–1533.

[84] Leslie WD, O'donnell S, Jean S, et al. Trends in hip fracture rates in Canada. *JAMA* 2009;**302**:883–9.

[85] Chen FP, Shyu YC, Fu TS, Sun CC, Chao AS, Tsai TL, Huang TS. Secular trends in incidence and recurrence rates of hip fracture: a nationwide population-based study. *Osteoporos Int* 2017;**28**:811–8.

[86] Xia WB, He SL, Xu L, Liu AM, Jiang Y, Li M, Wang O, Xing XP, Sun Y, Cummings SR. Rapidly increasing rates of hip fracture in Beijing, China. *J Bone Miner Res* 2012;**27**:125–9.

[87] Yoon HK, Park C, Jang S, Jang S, Lee YK, Ha YC. Incidence and mortality following hip fracture in Korea. *J Korean Med Sci* 2011;**26**:1087–92.

[88] Gibson-Smith D, Klop C, Elders PJ, Welsing PM, Van Schoor N, Leufkens HG, Harvey NC, van Staa TP, De Vries F. The risk of major and any (non-hip) fragility fracture after hip fracture in the United Kingdom: 2000-2010. *Osteoporos Int* 2014;**25**:2555–63.

[89] Melton 3rd LJ, Kearns AE, Atkinson EJ, Bolander ME, Achenbach SJ, Huddleston JM, Therneau TM, Leibson CL. Secular trends in hip fracture incidence and recurrence. *Osteoporos Int* 2009;**20**:687–94.

[90] Omsland TK, Emaus N, Tell GS, Magnus JH, Ahmed LA, Holvik K, Center J, Forsmo S, Gjesdal CG, Schei B, Vestergaard P, Eisman JA, Falch JA, Tverdal A, Søgaard AJ, Meyer HE. Mortality following the first hip fracture in Norwegian women and men (1999–2008). A NOREPOS study. *Bone* 2014;**63**:81–6.

[91] Bliuc D, Tran T, Alarkawi D, Nguyen TV, Eisman JA, Center JR. Secular changes in postfracture outcomes over 2 decades in Australia: a time-trend comparison of excess postfracture mortality in two birth controls over two decades. *J Clin Endocrinol Metab* 2016;**101**:2475–83.

[92] Melton 3rd LJ, Achenbach SJ, Atkinson EJ, Therneau TM, Amin S. Long-term mortality following fractures at different skeletal sites: a population-based cohort study. *Osteoporos Int* 2013;**24**:1689–96.

[93] Piirtola M, Vahlberg T, Lopponen M, Raiha I, Isoaho R, Kivela SL. Fractures as predictors of excess mortality in the aged-a population-based study with a 12-year follow-up. *Eur J Epidemiol* 2008;**23**:747–55.

[94] Shortt NL, Robinson CM. Mortality after low-energy fractures in patients aged at least 45 years old. *J Orthop Trauma* 2005;**19**:396–400.

[95] Bliuc D, Center JR. Determinants of mortality risk following osteoporotic fractures. *Curr Opin Rheumatol* 2016;**28**:413–9.

[96] Sattui SE, Saag KG. Fracture mortality: associations with epidemiology and osteoporosis treatment. *Nat Rev Endocrinol* 2014;**10**:592.

[97] Cauley JA, Thompson DE, Ensrud KC, Scott JC, Black D. Risk of mortality following clinical fractures. *Osteoporos Int* 2000;**11**:556–61.

[98] Center JR, Nguyen TV, Schneider D, Sambrook PN, Eisman JA. Mortality after all major types of osteoporotic fracture in men and women: an observational study. *Lancet* 1999;**353**:878–82.

[99] Tosteson AN, Gottlieb DJ, Radley DC, Fisher ES, Melton 3rd LJ. Excess mortality following hip fracture: the role of underlying health status. *Osteoporos Int* 2007;**18**:1463–72.

[100] Vestergaard P, Rejnmark L, Mosekilde L. Increased mortality in patients with a hip fracture-effect of pre-morbid conditions and post-fracture complications. *Osteoporos Int* 2007;**18**:1583–93.

[101] Bliuc D, Nguyen ND, Alarkawi D, Nguyen TV, Eisman JA, Center JR. Accelerated bone loss and increased post-fracture mortality in elderly women and men. *Osteoporos Int* 2015b;**26**:1331–9.

[102] Nguyen ND, Center JR, Eisman JA, Nguyen TV. Bone loss, weight loss, and weight fluctuation predict mortality risk in elderly men and women. *J Bone Miner Res* 2007a;**22**:1147–54.

[103] Rantanen T, Harris T, Leveille SG, Visser M, Foley D, Masaki K, Guralnik JM. Muscle strength and body mass index as long-term predictors of mortality in initially healthy men. *J Gerontol A Biol Sci Med Sci* 2000;**55**:M168–73.

[104] Sasaki H, Kasagi F, Yamada M, Fujita S. Grip strength predicts cause-specific mortality in middle-aged and elderly persons. *Am J Med* 2007;**120**:337–42.

[105] Chang SF, Lin PL. Systematic literature review and meta-analysis of the association of sarcopenia with mortality. *Worldviews Evid Based Nurs* 2016;**13**(2):153–62.

[106] Kannegaard PN, Van Der Mark S, Eiken P, Abrahamsen B. Excess mortality in men compared with women following a hip fracture. National analysis of comedications, comorbidity and survival. *Age Ageing* 2010;**39**:203–9.

[107] Nikkel LE, Kates SL, Schreck M, Maceroli M, Mahmood B, Elfar JC. Length of hospital stay after hip fracture and risk of early mortality after discharge in New York state: retrospective cohort study. *BMJ* 2015;**351**:h6246.

[108] Abrahamsen B, van Staa T, Ariely R, Olson M, Cooper C. Excess mortality following hip fracture: a systematic epidemiological review. *Osteoporos Int* 2009;**20**:1633–50.

[109] Kado DM, Browner WS, Palermo L, Nevitt MC, Genant HK, Cummings SR. Vertebral fractures and mortality in older women: a prospective study. Study of Osteoporotic Fractures Research Group. *Arch Intern Med* 1999;**159**:1215–20.

[110] Kado DM, Duong T, Stone KL, Ensrud KE, Nevitt MC, Greendale GA, Cummings SR. Incident vertebral fractures and mortality in older women: a prospective study. *Osteoporos Int* 2003;**14**:589–94.

[111] Ioannidis G, Papaioannou A, Hopman WM, Akhtar-Danesh N, Anastassiades T, Pickard L, Kennedy CC, Prior JC, Olszynski WP, Davison KS, Goltzman D, Thabane L, Gafni A, Papadimitropoulos EA, Brown JP, Josse RG, Hanley DA, Adachi JD. Relation between fractures and mortality: results from the Canadian Multicentre Osteoporosis Study. *CMAJ* 2009;**181**:265–71.

[112] Fink HA, Milavetz DL, Palermo L, Nevitt MC, Cauley JA, Genant HK, Black DM, Ensrud KE. What proportion of incident radiographic vertebral deformities is clinically diagnosed and vice versa? *J Bone Miner Res* 2005;**20**:1216–22.

[113] Ismail AA, O'neill TW, Cooper C, Finn JD, Bhalla AK, Cannata JB, Delmas P, Falch JA, Felsch B, Hoszowski K, Johnell O, Diaz-Lopez JB, Lopez Vaz A, Marchand F, Raspe H, Reid DM, Todd C, Weber K, Woolf A, Reeve J, Silman AJ. Mortality associated with vertebral deformity in men and women: results from the European Prospective Osteoporosis Study (EPOS). *Osteoporos Int* 1998;**8**:291–7.

[114] Trone DW, Kritz-Silverstein D, Von Muhlen DG, Wingard DL, Barrett-Connor E. Is radiographic vertebral fracture a risk factor for mortality? *Am J Epidemiol* 2007;**166**:1191–7.

[115] Hasserius R, Karlsson MK, Jonsson B, Redlund-Johnell I, Johnell O. Long-term morbidity and mortality after a clinically diagnosed vertebral fracture in the elderly–a 12- and 22-year follow-up of 257 patients. *Calcif Tissue Int* 2005;**76**:235–42. Epub 2005 Apr 11.

[116] Huntjens KM, Kosar S, Van Geel TA, Geusens PP, Willems P, Kessels A, Winkens B, Brink P, Van Helden S. Risk of subsequent fracture and mortality within 5 years after a non-vertebral fracture. *Osteoporos Int* 2010;**21**:2075–82.

[117] Tran T, Bliuc D, Van Geel T, Adachi JD, Berger C, Van Den Bergh J, Eisman JA, Geusens P, Goltzman D, Hanley DA, Josse RG, Kaiser SM, Kovacs CS, Langsetmo L, Prior JC, Nguyen TV, Center JR. Population-wide impact of non-hip non-vertebral fractures on mortality. *J Bone Miner Res* 2017;**32**(9):1802–10.

[118] Morin S, Lix LM, Azimaee M, Metge C, Caetano P, Leslie WD. Mortality rates after incident non-traumatic fractures in older men and women. *Osteoporos Int* 2011;**22**:2439–48.

[119] Somersalo A, Paloneva J, Kautiainen H, Lonnroos E, Heinanen M, Kiviranta I. Increased mortality after upper extremity fracture requiring inpatient care. *Acta Orthop* 2015;**86**:533–57.

[120] Øyen J, Diamantopoulos AP, Haugeberg G. Mortality after distal radius fracture in men and women aged 50 years and older in Southern Norway. *PLoS One* 2014;**9**:e112098.

[121] Bouza C, Lopez T, Palma M, Amate JM. Hospitalised osteoporotic vertebral fractures in Spain: analysis of the national hospital discharge registry. *Osteoporos Int* 2007;**18**:649–57.

[122] Klop C, Welsing PM, Cooper C, Harvey NC, Elders PJ, Bijlsma JW, Leufkens HG, De Vries F. Mortality in British hip fracture patients, 2000-2010: a population-based retrospective cohort study. *Bone* 2014;**66**:171–7.

[123] Maravic M, Taupin P, Landais P, Roux C. Decrease of inpatient mortality for hip fracture in France. *Joint Bone Spine* 2011;**78**:506–9.

[124] Orces CH. In-hospital hip fracture mortality trends in older adults: the National Hospital Discharge Survey, 1988-2007. *J Am Geriatr Soc* 2013;**61**:2248–9.

[125] Wu TY, Jen MH, Bottle A, Liaw CK, Aylin P, Majeed A. Admission rates and in-hospital mortality for hip fractures in England 1998 to 2009: time trends study. *J Public Health (Oxf)* 2011;**33**:284–91.

[126] Haleem S, Lutchman L, Mayahi R, Grice JE, Parker MJ. Mortality following hip fracture: trends and geographical variations over the last 40 years. *Injury* 2008;**39**:1157–63.

[127] Karampampa K, Ahlbom A, Michaelsson K, Andersson T, Drefahl S, Modig K. Declining incidence trends for hip fractures have not been accompanied by improvements in lifetime risk or post-fracture survival–A nationwide study of the Swedish population 60 years and older. *Bone* 2015;**78**:55–61.

[128] Chen JS, Sambrook PN. Antiresorptive therapies for osteoporosis: a clinical overview. *Nat Rev Endocrinol* 2012;**8**:81–91.

[129] Reginster JY. Antifracture efficacy of currently available therapies for postmenopausal osteoporosis. *Drugs* 2011;**71**:65–78.

[130] Chen L, Wang G, Zheng F, Zhao H, Li H. Efficacy of bisphosphonates against osteoporosis in adult men: a meta-analysis of randomized controlled trials. *Osteoporos Int* 2015;**26**:2355–63.

[131] Hochberg MC. Nonvertebral fracture risk reduction with nitrogen-containing bisphosphonates. *Curr Osteoporos Rep* 2008;**6**:89–94.

[132] Liu M, Guo L, Pei Y, Li N, Jin M, Ma L, Liu Y, Sun B, Li C. Efficacy of zoledronic acid in treatment of osteoporosis in men and women-a meta-analysis. *Int J Clin Exp Med* 2015;**8**:3855–61.

[133] Bolland MJ, Grey AB, Gamble GD, Reid IR. Effect of osteoporosis treatment on mortality: a meta-analysis. *J Clin Endocrinol Metab* 2010;**95**:1174–81.

[134] Lyles KW, Colon-Emeric CS, Magaziner JS, Adachi JD, Pieper CF, Mautalen C, Hyldstrup L, Recknor C, Nordsletten L, Moore KA, Lavecchia C, Zhang J, Mesenbrink P, Hodgson PK, Abrams K, Orloff JJ, Horowitz Z, Eriksen EF, Boonen S. Zoledronic acid and clinical fractures and mortality after hip fracture. *N Engl J Med* 2007;**357**:1799–809.

[135] Beaupre LA, Morrish DW, Hanley DA, Maksymowych WP, Bell NR, Juby AG, Majumdar SR. Oral bisphosphonates are associated with reduced mortality after hip fracture. *Osteoporos Int* 2010;**22**(3):983–91.

[136] Bondo L, Eiken P, Abrahamsen B. Analysis of the association between bisphosphonate treatment survival in Danish hip fracture patients-a nationwide register-based open cohort study. *Osteoporos Int* 2013;**24**:245–52.

[137] Center JR, Bliuc D, Nguyen ND, Nguyen TV, Eisman JA. Osteoporosis medication and reduced mortality risk in elderly women and men. *J Clin Endocrinol Metab* 2011;**96**(4):1006–14.

[138] Cree MW, Juby AG, Carriere KC. Mortality and morbidity associated with osteoporosis drug treatment following hip fracture. *Osteoporos Int* 2003;**14**:722–7.

[139] Sambrook PN, Cameron ID, Chen JS, March LM, Simpson JM, Cumming RG, Seibel MJ. Oral bisphosphonates are associated with reduced mortality in frail older people: a prospective five-year study. *Osteoporos Int* 2010;**22**(9):2551–6.

[140] Lee P, Ng C, Slattery A, Nair P, Eisman JA, Center JR. Preadmission bisphosphonate and mortality in critically ill patients. *J Clin Endocrinol Metab* 2016;**101**(5):1945–53. jc20153467.

[141] Brozek W, Reichardt B, Zwerina J, Dimai HP, Klaushofer K, Zwettler E. Antiresorptive therapy and risk of mortality and refracture in osteoporosis-related hip fracture: a nationwide study. *Osteoporos Int* 2016;**27**:387–96.

[142] Nurmi-Luthje I, Luthje P, Kaukonen JP, Kataja M, Kuurne S, Naboulsi H, Karjalainen K. Post-fracture prescribed calcium and vitamin D supplements alone or, in females, with concomitant anti-osteoporotic drugs is associated with lower mortality in elderly hip fracture patients: a prospective analysis. *Drugs Aging* 2009;**26**:409–21.

[143] Chen YC, Su FM, Cheng TT, Lin WC, Lui CC. Can antiosteoporotic therapy reduce mortality in MRI-proved acute osteoporotic vertebral fractures? *J Bone Miner Metab* 2016;**34**:325–30.

[144] Colon-Emeric CS, Mesenbrink P, Lyles KW, Pieper CF, Boonen S, Delmas P, Eriksen EF, Magaziner J. Potential mediators of the mortality reduction with zoledronic acid after hip fracture. *J Bone Miner Res* 2010;**25**:91–7.

[145] Kado DM, Browner WS, Blackwell T, Gore R, Cummings SR. Rate of bone loss is associated with mortality in older women: a prospective study. *J Bone Miner Res* 2000;**15**:1974–80.

[146] Cauley JA, Lui LY, Barnes D, Ensrud KE, Zmuda JM, Hillier TA, Hochberg MC, Schwartz AV, Yaffe K, Cummings SR, Newman AB. Successful skeletal aging: a marker of low fracture risk and longevity. The Study of Osteoporotic Fractures (SOF). *J Bone Miner Res* 2009;**24**:134–43.

[147] Corrado A, Santoro N, Cantatore FP. Extra-skeletal effects of bisphosphonates. *Joint Bone Spine* 2007;**74**:32–8.

[148] Toussirot E, Wendling D. Antiinflammatory treatment with bisphosphonates in ankylosing spondylitis. *Curr Opin Rheumatol* 2007;**19**:340–5.

Chapter 3

Models of Secondary Fracture Prevention: Systematic Review and Metaanalysis of Outcomes

Kirtan Ganda[1,2], Paul J. Mitchell[3,4], Markus J. Seibel[1,2,5]
[1]Department of Endocrinology and Metabolism, Concord Repatriation General Hospital, Concord, NSW, Australia; [2]Concord Clinical School, University of Sydney, Sydney, NSW, Australia; [3]School of Medicine, Sydney Campus, University of Notre Dame Australia, Sydney, NSW, Australia; [4]Osteoporosis New Zealand, Wellington, New Zealand; [5]Bone Research Program, ANZAC Research Institute, Sydney, NSW, Australia

SECONDARY FRACTURE PREVENTION PROGRAMMES

After a symptomatic fracture occurs, patients are typically assessed in the emergency department of a hospital. The patient then comes into contact with an orthopaedic service, either seen as an outpatient or inpatient. This represents an ideal window of opportunity to capture and investigate patients to establish whether their fracture is due to bone fragility and requires therapy to prevent further fractures.

Over the last 15 years, a substantial body of literature from across the world has been created in relation to the design and clinical efficacy of Secondary Fracture Prevention (SFP) programmes (or 'Fracture Liaison Services', FLSs). However, depending on the health-care system and available resources, the degree of intervention can vary greatly between programmes. Thus, the published literature on SFP programmes is characterized by significant heterogeneity in regards to the methods employed for patient identification and, in particular, the intensity of the intervention.

Types of intervention vary across a wide spectrum of approaches, ranging from a simple, education-based model that has a high rate of patient turnover and requiring few resources to more complex models involving a dedicated service coordinator who identifies and then channels patients with suspected fragility fractures into a specific programme to investigate and, as appropriate, manage patients with osteoporotic fractures. These complex and intensive models typically incorporate patient education and risk assessment, with on-site DXA scanning, as well as treatment initiation within one care package. The coordinator usually plays a pivotal role in orchestrating these steps.

Secondary Fracture Prevention. https://doi.org/10.1016/B978-0-12-813136-7.00003-X

TABLE 3.1 Models of Care, or Intervention Types for SFP programmes

Intervention Type	Description
A	Identification, assessment (risk factors, bloods, BMD), treatment initiation
B	Identification, assessment, treatment recommendation only
C	Education of patient and primary care physician
D	Education of patient only

Given the heterogeneity in type and degree of interventions, SFP programmes have been classified in 2012 by Ganda et al. into different categories graded from A through to D, with model A representing the most intensive, all-encompassing intervention, and model D being the least intensive.[1] Although each model of care is unique, there are significant variations in details of the intervention as well as reporting of outcomes between each study (Table 3.1).

SEARCH METHODOLOGY

The present chapter is a contemporary update of a systematic review undertaken in 2012 by Ganda et al.[1] It aims to summarize the design and clinical outcomes of different SFP models from around the world, published up to mid-2017. Medline, Pubmed, Cochrane and Embase databases were searched using key words singularly and in various combinations. These key words were 'osteoporosis, fracture, strategy/-ies, intervention/-s, programme/-s, prevention, implementation, identification, minimal/low/fragility fracture, fracture liaison services, secondary fracture prevention programme/-s'. The searches were limited from 1996 to June 2017, English language articles and patients aged 45 years and older. Articles relating to primary fracture prevention were excluded. Data extracted from each article included the study design, the type of intervention (i.e., A, B, C or D) as described in 2012 by Ganda et al. and specific outcome measures. The outcome measures extracted were rates of bone mineral density (BMD) testing, rates of osteoporosis pharmacotherapy initiation, adherence to therapy, refracture rates and cost-effectiveness analyses.

The studies with valid control groups for the outcomes of BMD testing and treatment initiation rates were then included in a metaanalysis stratified by the intervention type or 'model of care' (i.e., type A, B and C). The aim of the analysis was to determine the most effective model of care and to guide the future development of SFP programmes. Type D interventions were not included in the metaanalysis as there was insufficient data.

Type A Model Interventions

Since 2000, a total of 34 studies describing type A SFP programmes have been published. All of these reports describe a coordinated model of care, which identifies, assesses and treats patients after their minimal trauma fracture, as part of an all-encompassing clinical service. Assessments and investigations include evaluation of clinical risk factors for osteoporosis, a BMD scan and pathology tests to exclude secondary causes for osteoporosis. In patients with confirmed osteoporosis, lifestyle and dietary advice, falls prevention as well as pharmacological interventions are initiated within the programme, as appropriate.

A typical example of a type A model of care is the Concord Hospital SFP programme, a service run by a dedicated, medically trained coordinator (Fig. 3.1). Other examples of type A models can be found in the United States, in particular where health care is provided by large private insurance companies or other health maintenance organisations (HMO) with fully integrated electronic patient records. The Kaiser Permanente Group in Southern California is an HMO with 3.1 million members, spanning across eleven medical centres. Their 'Healthy Bones Programme' consists of a highly organized process to identify high-risk patients, followed by a complete clinical assessment and, if indicated, treatment for osteoporosis. The programme is assisted by a fully integrated electronic health record, allowing for a 'closed system' approach. Thus, these organisations are able to monitor patient outcomes on a monthly basis via reports for 'care managers', detailing those who needed treatment and those who received treatment, thereby providing targets and benchmarks, which stimulate further improvements in care.

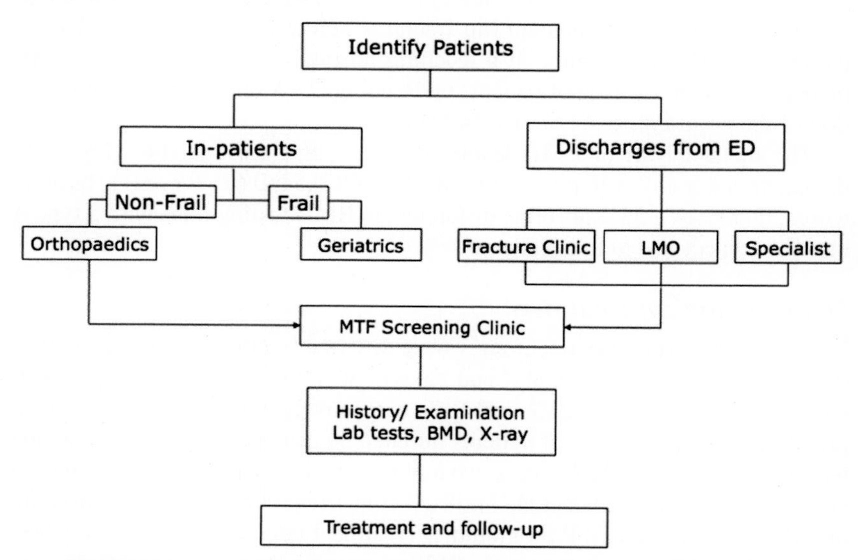

FIGURE 3.1 A prototype of a type A model Secondary Fracture Prevention program.

Outcomes

Amongst the 34 studies describing a type A model of care, 13 have been included in the metaanalysis due to the presence of valid control groups (Table 3.2). The 11 studies, which included adherence data with or without control groups are described below in narrative style. There were seven studies with refracture data (vs. controls) and five studies describing cost-effectiveness of this model of care.

Bone Mineral Density Testing

Vaile et al.,[2] Kuo et al.[3] and Yates et al.[4] utilised historic control groups, in which BMD testing rates were as low as 20%, 26% and 28%, respectively. In contrast, BMD scans were performed in 100%, 83% and 100% of patients managed with the SFP programmes.

An RCT by Majumdar et al.[5] reported BMD testing rates of 29% (32/110) in the control group and of 80% in the intervention group at 6 months posthip fracture. In another RCT by the same group,[6] BMD testing rates 6 months postwrist fracture were 52% in the control arm and 81% in the intervention arm.

A cross-sectional study in 2006 by Streeten et al.[7] demonstrated a 51% DXA scanning rate in the intervention arm and no patients being referred for a BMD scan in the control arm.

In 2013, Queally et al.[8] published a small prospective randomized trial comparing a type A with a type C model of care. As expected, the proportion of patients who underwent BMD testing was significantly greater in the type A than the type C model (86% vs. 36%).

In 2015, Ruggiero et al., an Italian group published a paper describing outcomes of patients with incident hip fractures before and after the implementation of an SFP programme at a teaching hospital.[9] There was a significant increase in the percentage of patients undergoing DXA scans after implementation of the programme (47.6% vs. 14.5%).

The metaanalysis of BMD testing rates amongst these studies (Fig. 3.2), demonstrated a risk difference of 0.58 (95% CI 0.54–0.62, $P<.001$). In other words, there was a 58% absolute difference in BMD testing rates with a type A intervention compared to usual care.

Pharmacological Treatment

Specific osteoporosis treatment rates were described in three 'before and after' studies (Vaile et al.,[2] Kuo et al.,[3] and Yates et al.[4]). In the control groups, treatment rates were of 11%, 21% and 10%, respectively, whereas within the SFP programme the proportion of patients initiated on osteoporosis-specific therapy was of 34%, 28% and 22% (i.e., a two to threefold increase). Two studies by Lih et al.[10] and Van der Kallen et al.[11] utilised concurrent control groups consisting of nonattendees to the SFP programme. Treatment rates were of 32% and 38% in the control groups, whereas in the SFP programme the corresponding proportion of treated patients were 80% and 71%.

TABLE 3.2 Summary of Studies Included in Metaanalysis of Intervention Type A Studies

<table>
<tr><th>Study Name</th><th>Study Type</th><th>Fracture Site</th><th>Age</th><th>Female (%)</th><th>BMD (Control)</th><th>BMD (Intervention)</th><th>Treatment (Control)</th><th>Treatment (Intervention)</th></tr>
<tr><td>Vaile et al.[2]</td><td>Before & After</td><td>All (nil breakdown)</td><td>>55</td><td>–</td><td>31|157</td><td>983|983</td><td>17|157</td><td>334|983</td></tr>
<tr><td>Lih et al.[10]</td><td>Prospective Controlled</td><td>All (35% wrist)</td><td>66 (mean)</td><td>80</td><td>–</td><td>–</td><td>51|157</td><td>198|246</td></tr>
<tr><td>Kuo et al.[3]</td><td>Before & After</td><td>All (I only reported)</td><td>64 (mean)</td><td>71</td><td>40|155</td><td>95|115</td><td>32|155</td><td>35|123</td></tr>
<tr><td>Jones et al.[53]</td><td>Before & After</td><td>Hip (NOF)</td><td>81 (mean)</td><td>72</td><td>–</td><td>–</td><td>8|161</td><td>22|93</td></tr>
<tr><td>Streeten et al.[7]</td><td>Cross-Sectional</td><td>Hip (C – 100%);
Hip (I – 53%)</td><td>70 (mean)</td><td>46</td><td>0|31</td><td>27|53</td><td>1|31</td><td>28|53</td></tr>
<tr><td>Edwards et al.[12]</td><td>Before & After</td><td>All</td><td>73 (mean)</td><td>82</td><td>–</td><td>–</td><td>14|38</td><td>93|151</td></tr>
<tr><td>Majumdar et al.[6]</td><td>RCT</td><td>Wrist</td><td>60 (median)</td><td>68</td><td>13|25</td><td>17|21</td><td>3|25</td><td>9|21</td></tr>
<tr><td>Majumdar et al.[5]</td><td>RCT</td><td>Hip</td><td>75.9 (median)</td><td>65</td><td>32|110</td><td>88|110</td><td>24|110</td><td>56|110</td></tr>
<tr><td>Olenginski et al.[13]</td><td>Retrospective audit</td><td>All (Hip 58%)</td><td>77.6 (mean)</td><td>79</td><td>–</td><td>–</td><td>29|90</td><td>308|382</td></tr>
<tr><td>Queally et al.[8]</td><td>RCT</td><td>All</td><td>64 (mean)</td><td>70</td><td>11|30</td><td>21|31</td><td>5|30</td><td>12|31</td></tr>
<tr><td>Ruggiero et al.[9]</td><td>Before & After</td><td>Hip</td><td>83 (mean)</td><td>75</td><td>25|172</td><td>100|210</td><td>29|172</td><td>98|210</td></tr>
<tr><td>Van der Kallen et al.[11]</td><td>Prospective controlled (controlled = nonattendees)</td><td>All</td><td>73 (mean)</td><td>78</td><td>–</td><td>–</td><td>82|214</td><td>156|220</td></tr>
<tr><td>Yates et al.[4]</td><td>Before & After</td><td>All</td><td>67 (mean)</td><td>77</td><td>16|58</td><td>203|203</td><td>6|58</td><td>45|203</td></tr>
</table>

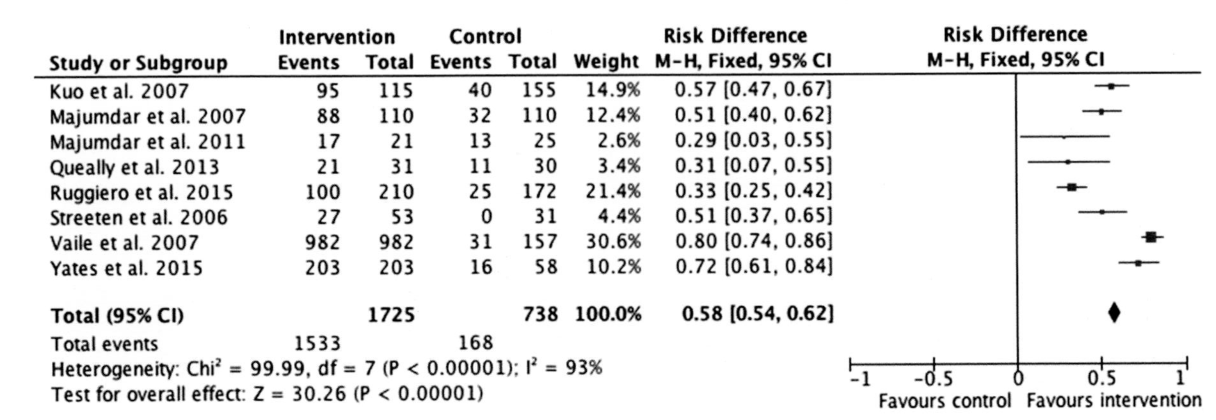

FIGURE 3.2 Metaanalysis of BMD testing rates using risk difference in intervention type A studies.

In 2005, Edwards et al.[12] implemented an SFP programme at the McGraw Medical Centre in Chicago. Amongst 151 patients seen in the programme, 19% (25/151) were on therapy prior to the index fracture (2/3 of which were hip fractures). After 6 months, 62% (93/151) were on pharmacotherapy for osteoporosis.

In 2015, Olenginski et al. published a retrospective audit of a type A versus type C model of care at Geisinger Health (an HMO in the United States).[13] The rates of treatment initiation in the type A model was 81% (308/382) compared to 32% (29/90) in the type C model of care.

An RCT in 2007 by Majumdar et al.[5] reported that 51% (56/110) of the intervention group were taking bisphosphonate therapy 6 months posthip fracture, compared with 22% (24/110) in the control group. Treatment rates in the later but rather small study in 2011 by Majumdar et al.[6] were 43% (9/21) in the intervention group and 12% (3/25) in the control group at 6 months. Similarly, a small study in 2013 by Queally et al.[8] reported that 12 of 31 patients (39%) seen in a type A service were initiated on treatment, compared with 5 out of 30 patients (17%) seen in type C clinic. Initiation rates for osteoporosis pharmacotherapy reported by Ruggiero et al.[9] were 48.5% for the intervention arm and 17.2% for care as usual.

Metaanalysis of treatment initiation rates amongst these studies (Fig. 3.3), demonstrated a risk difference of 0.29 (95% CI 0.26–0.32, $P < .001$), indicating a 29% absolute difference in treatment initiation rates with a type A intervention compared with usual care or type C models of care.

Adherence

Adherence to osteoporosis pharmacotherapy has been reported in 11 studies. In 2014, Ganda et al. compared compliance and persistence to oral bisphosphonate therapy in a randomized controlled study design amongst patients initiated on therapy by the Concord SFP programme.[14] Patients were randomized to either six-monthly follow-up by the SFP programme or management by their primary care physician. Compliance and persistence rates at 2 years postrandomization were determined utilizing pharmaceutical claims and were found to be similar in both groups (compliance: 78%, persistence: 62% at 2 years). These results demonstrated that once therapy has been initiated by the SFP programme, primary care physicians are entirely capable of maintaining patient adherence to therapy. Thus, it appears that the osteoporosis treatment gap is partly due to primary care physicians being reluctant to initiate therapy.

In 2014, Van der Kallen et al.[11] described a self-reported adherence of 67% after a mean follow-up of 22 months amongst the attendees of the John Hunter Hospital SFP programme in Newcastle, compared with 34% in nonattendees. In 2015, Yates et al.[4] from the Royal Melbourne Hospital demonstrated a self-reported adherence of 84% at 2 years. In 2007, Vaile et al. described self-reported adherence of 95% after 12 months at the SFP programme in Sydney's Royal Prince Alfred Hospital amongst patients who attended the 12-month

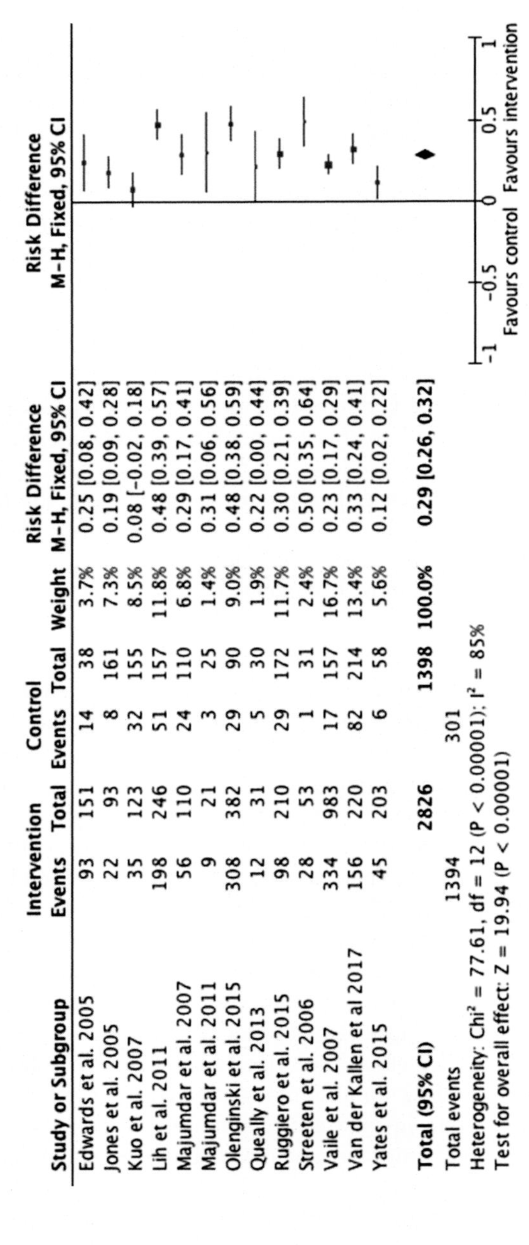

FIGURE 3.3 Metaanalysis of treatment initiation rates using risk difference in intervention type A studies.

follow-up visit.[2] In 2007, Kuo et al.[3] from St. Vincent's Hospital in Sydney reported 80% compliance with therapy initiated by the SFP programme after a mean follow-up of 10 months.

In 2015, the Italian group of Ruggiero et al.[9] published a paper describing outcomes of patients with incident hip fractures before and after the implementation of a type A SFP programme at a teaching hospital. Self-reported persistence at 12 months was 57% (56/98) in the SFP group and 59% (17/29) in the historical control group.

In 2014, Dehamchi-Rehailia et al.[15] from Amiens University Hospital, France identified 872 patients with a minimal trauma fracture over 2 years, with 335 patients captured by the SFP programme. Treatment was initiated in 54% of patients (182/335). Seventy-four percentage (n=123) of 166 patients with follow-up data available were persistent with therapy after 12 months.

In 2015, Naranjo et al. published a prospective study from the North Grand Canary Island health area in Spain reported data from a type A SFP programme.[16] During 2 years, 1674 patients presented with a minimal trauma fracture were identified, of whom 759 were seen in the programme. All patients who attended the SFP programme had DXA scans performed. Treatment was initiated in 72% of patients (n=549) and persistence at 2 years was 73% (85/116). However, 79% of the patients initially seen by the programme had dropped out by 2 years, indicating the presence of a strong positive bias. Persistence was self-reported and confirmed using a prescription database. Amongst a sample of those who dropped out, the majority were nonadherent to therapy, indicated that persistence was much lower amongst those who did not attend follow-up.

In 2005, Edwards et al.[12] described a treatment initiation rate after 6 months of 62% (93/151) but only 21% (32/151) remained on therapy after 12 months. This illustrates the pivotal role of primary care physicians in the long-term follow-up of these patients and the need to make both patients and family doctors aware of the importance of long-term, sustained therapy for osteoporosis.

In 2013, Goltz et al. published a case-control study from Dresden, Germany, used a health insurance data set (rather than self-reported adherence) to compare outcomes from 2455 patients with minimal trauma fractures enrolled in a type A SFP programme with the same number of matched controls who did not participate in the programme.[17] Further fracture incidence over 3 years was determined using ICD-10 coding data. However, there was no significant difference in refracture rates between the intervention and control arms (9.7% vs. 10.0%). Data over 3 years (2007–10) indicated average compliance with osteoporosis pharmacotherapy of 80% or more in 59% (157/269) in the intervention group versus 37% (99/269) in the control group. Persistence rates were not reported.

In 2011, Boudou et al.[18] reported data from a type A SFP programme at the University Hospital of Saint Etienne, France. The study identified 1691 patients with a minimal trauma fracture over 3.5 years, with 279 captured or seen in the hospital's SFP programme. Amongst 155 patients with complete follow-up data, 90% (n=140) commenced therapy, mostly in the form of oral weekly bisphosphonates. Self-reported data from 140 patients revealed that persistence to therapy was high at 89% (124/140) after 12 months and 75% (105/140) after a mean follow-up of 29 months.

In 2013, Eekman[19] described data from four SFP programmes in the Netherlands. The majority of patients were initiated on oral bisphosphonates as first line therapy. Amongst those who attended the SFP programme (51% of those invited – 1116/2207), DXA was performed in 100% of patients. Amongst the 337 patients that were initiated on treatment, there were 280 (83%) with follow-up data at 1 year. Self-reported persistence was 88% (N=246) and 2% (N=6) sustained further fractures.

Refractures

Refracture rates with control groups have been reported in five studies. In 2001, Lih et al.[10] reported an 80% relative risk reduction amongst those attending versus nonattenders to the Concord Hospital SFP programme after 4 years follow-up (4.1% SFP group vs. 19.7% nonattenders). In a similar study design, in 2014 Van der Kallen et al.[11] reported a 65% relative risk reduction (18.6% vs. 6.5%) after approximately 2 years of follow-up. These risk reductions may be an overestimate as the attenders may have been individuals more likely to be adherent. However, the characteristics of the groups seemed to be well matched in both studies. In 2015, Ganda et al.[20] from the Concord Hospital SFP programme published an observational study with up to 7 years of follow-up, with the aim of determining the predictors of refracture amongst those managed in the SFP programme. The predictors of refracture included multimorbidity, low hip BMD, poor compliance to therapy and corticosteroid use. The most recent analysis of refractures in 2016 by Nakayama et al.[21] compared refracture rates in a hospital with an existing SFP programme to a similar sized hospital without such a programme. After 3 years of follow-up the relative risk reduction in refractures was about 30% (12% SFP vs. 17% non-SFP). In 2014, Huntjens et al.[22] compared nonvertebral refracture rates in patients who attended an SFP programme after their first fracture (n=1412) versus those who did not (n=1910). Although refracture rates at 24 months were similar in both groups (6.7% vs. 6.8%), multivariate Cox regression demonstrated a reduction in refractures in patients who attended the SFP programme (HR 0.44; 95% CI 0.25–0.79). In addition, mortality was reduced by 35% in those managed by the SFP programme (HR 0.65; 95% CI 0.53–0.79), a number not dissimilar from the reduction in mortality seen in patients treated with IV bisphosphonates after a hip fracture.[23]

In 2008, Dell et al.[24] described the effect of the Kaiser Permanente SFP programme on refracture rates. Of note, there was no active control group. Instead, the authors used historical data from 1997–99 (pre-SFP programme) to estimate expected rates of hip fracture, and compared these to the rate of hip fractures observed in 2006, after the introduction of the SFP programme. Overall, there was a 37% reduction in the observed versus expected hip fractures incidence.

Cost-Effectiveness

There have been five studies describing the cost-effectiveness of type A intervention models. An informal evaluation of cost-effectiveness was described by Vaile et al.,[2] who estimated that if one hip fracture is prevented, the saving of AUD23,000 could pay for half the annual salary of an SFP programme coordinator, or for the clinical assessment of 54 patients with minimal trauma fractures.

A cost-effectiveness analysis of the Concord Hospital SFP programme,[25] revealed a cost-effectiveness ratio versus standard care of AUD17,291 per Quality Adjusted Life Year (QALY) gained. In 2015, Yates et al.[4] demonstrated an incremental cost-effectiveness ratio of AUD31,740 per QALY gained when compared to a type C intervention. Thus, the Concord Hospital and Royal Melbourne Hospital SFP programmes have been shown to be cost-effective interventions.

In 2008, Sander et al.[26] performed a cost-effectiveness analysis of the Toronto-based SFP programme, with data derived from a previous study by Bogoch et al.[27] Compared to usual care, a type A SFP programme was estimated to reduce hip fractures from 34 to 31 over a year, if the service attends to 500 patients per year. The cost saving was estimated to be C$49,950 (Canadian dollars in year-2004 values).

In 2009, Majumdar et al.[28] performed a cost-effectiveness analysis based on the RCT that included hip fracture patients. There was an estimated incremental cost saving of C$2576 for every 100 patients seen by the SFP programme. Moreover, there would be six fractures prevented, four quality-adjusted life years gained and C$260,000 saved (Canadian dollars in year-2006 values).

Type B Model Interventions

Unlike in a type A model of care, treatment initiation in a type B intervention model is the responsibility of the primary care physician. The United Kingdom has seen a concerted effort towards refracture prevention, attempting to establish SFP programmes in all hospitals.[29] These SFP programmes are usually based on the Glasgow model and are a typical example of a type B model of care.

The strong advocacy towards refracture prevention has been supported and catalysed by patient and professional organisations and the government. The latter has introduced financial incentives to primary care physicians who follow the guidelines, resulting in relatively good adherence to treatment recommendations issued by the SFP programmes. In 2007, the National Hip Fracture Database (NHFD) was established, which allows benchmarking of hip fracture care across the United Kingdom, thereby catalysing improvements in clinical care. As a testimony to the success of the programme, all 200 hospitals in England, Wales and Northern Ireland have subscribed to the database. In addition, the most recent report from the NHFD in 2017, describes 60% of hip fracture patients being offered osteoporosis-specific treatment on discharge from hospital (i.e., an antiresorptive or anabolic agent, not calcium and/or vitamin D alone).

Outcomes

There have been 15 studies describing type B interventions; however, only studies with valid control group (N=8) were included in the metaanalysis of BMD and treatment initiation rates (Table 3.3). Three studies reported adherence data and two reports performed cost-effectiveness analyses. No studies reported on refracture rates.

Bone Mineral Density Testing and Treatment Initiation Rates

In 2006, Bluic et al.[30] performed a randomised controlled trial, comparing a type B intervention with a type D intervention amongst 80 and 79 patients, respectively. The type B intervention involved a letter with an offer for a free BMD test, whereas the type D intervention issued patients with a letter only. The rate of BMD testing for type B versus D interventions was 38% (30/79) versus 7% (5/75), respectively, whereas the rate of treatment initiation was similar at 5% (4/79) versus 7% (5/75), respectively.

The 'before and after' study in 2004 by Cuddihy et al.[31] included all residents of Olmsted County, Minnesota, USA, with distal forearm fractures. Amongst the 105 eligible patients identified, 59 (56%) consented to participate in the intervention. Amongst those who consented to participate, 71% (42/59) were tested for BMD whereas prior to the intervention, only 5% (17/343) of patients had a BMD scan performed. Thirty-eight patients (38/59 or 64%) either continued or initiated therapy after the intervention. Prior to the intervention, only 17% (58/343) of patients were recommended for osteoporosis pharmacotherapy.

In 2005, Harrington et al.[32] from the University of Wisconsin Medical Foundation, USA, published outcome data from an SFP programme involving gradual intensification of the intervention in progressive phases, i.e., cycle 1 through to cycle 3. Cycle 2 involved a comparison of a type B and 'usual care' over the same 5-month period and provided the clearest data and thus is described here. In cycle 2, three orthopaedists agreed to the intervention (type B), whereas four orthopaedists did not agree ('usual care group'). Amongst 37 of 42 eligible patients who agreed to participate, 27 (27/42 or 64%) had DXA scans obtained, although 4 attended the primary care physician rather than SFP programme for assessment. None of the 55 patients in the usual care group had a DXA scan. Treatment was initiated in 25 of 37 patients (68%) in the intervention group as compared to 3/55 (5%) in the usual care group.

In 2005, Johnson et al.[33] published and pre- and postintervention study based at McGuire Veterans Affairs Medical Center, Richmond, Virginia, USA. As this was a Veteran's facility, 95% of patients in the study were male. The intervention, a type B SFP programme was compared to usual care preintervention. Patients were identified over a 6-month period in both groups. The results were reported as intention-to-treat, which meant the denominator was the number of patients identified as eligible, rather than the number seen in the programme or captured. The rate of BMD testing was 64% (85/136) in the intervention group as compared to 8% (16/126) in the preintervention group. The rates of specific

TABLE 3.3 Summary of Studies Included in Metaanalysis of Intervention Type B Studies

Study Name	Study Type	Fracture Site	Age	Female (%)	BMD (Control)	BMD (Intervention)	Treatment (Control)	Treatment (Intervention)
Bliuc et al.[30]	RCT	All	52.7 (mean)	50	5|75	30|79	5|75	4|79
Sidwell et al.[36]	Before & After	All (mostly hip 133/193; control 101/178 I)	81 (mean)	75	20|178	158|193	16|178	40|193
Cuddihy et al.[31]	Before & After	Wrist	68 (mean)	86	17|343	42|59	55|343	17|30
Johnson et al.[33]	Before & After	All	59 (mean)	4	16|126	85|103	15|126	57|136
Harrington et al.[32] (cycle 2)	Prospective Controlled	All	–	–	0|55	27|37	3|55	22|37
Morrish et al.[35]	RCT	Hip	75.9 (median)	65	32|110	75|110	24|110	42|110
Wallace et al.[37]	Cross-Sectional	Hip	84 (median)	100	0|46	1|42	28|46	38|42
Lee et al.[34]	Cross-Sectional	All	70 (mean)	6	75|203	74|118	–	–

osteoporosis pharmacotherapy were 20% (32/136) in the intervention group and 5% (12/126) in the preintervention group.

The most recent publication from the United States by Lee et al., in 2016[34] described outcomes from a network of Veterans Affairs Medical Centers, which consisted primarily of male patients (95%). This SFP programme utilised an electronic consult (e-consult) by a bone specialist which was sent to primary care physicians with a view to improving osteoporosis management, particularly amongst the rural population of veterans, as specialty access is limited in these areas. During 1 year, 321 e-consults were performed, 53% (n = 171) of which were for patients from rural areas. A Fracture Liaison Coordinator ordered BMD testing and other investigations for 118 patients (type B intervention), whereas the remaining 203 patients did not have coordinator involvement (type C intervention). A unique outcome measure reported in this study was 19,178 km of travelling was saved for rural patients. The type B intervention resulted in BMD testing rate of 63% (74/118) whereas the type C intervention had a rate of 37% (75/203). Amongst those recommended treatment, initiation occurred in 76% in the type B intervention and in 40% of patients in the type C intervention.

In 2009, Morrish et al.[35] published an extension of the RCT by Majumdar et al.[5] conducted in Alberta, Canada. The original trial (described above) compared a type A intervention with usual care, with 110 patients in each group over a 6-month period. The authors assigned the usual group to a type B intervention at 6 months and reported the outcomes after a further 6 months of follow-up. Bone density testing rates increased with increasing intensity of intervention. That is, the rates of BMD testing amongst those undergoing usual care, type B intervention and type A intervention were 29% (32/110), 68% (75/110) and 80% (88/110), respectively. Similarly, rates of osteoporosis pharmacotherapy for usual care, type B and type A interventions were 22% (24/110), 38% (42/110) and 54% (59/110), respectively. There was no significant difference in the number of refractures at 12 months in the type A versus B interventions.

In 2004, Sidwell et al.[36] reported results before and after implementing a type B SFP programme at Princess Margaret Hospital, Christchurch, New Zealand. Amongst 193 patients in the intervention group, 150 underwent a BMD scan (78%), whereas 20 of 178 patients (11%) had a BMD scan in the preintervention group. Osteoporosis pharmacotherapy was initiated in 40 of 193 patients (21%) in the intervention group and 16 of 178 (9%) patients in the preintervention group.

In 2011, Wallace et al.[37] in a cross-sectional study compared a service with a type B SFP programme to a service without an SFP programme in a population of female patients with hip fractures. There was little difference in the rate of BMD testing between the two groups, which is consistent with the fact that these were high-risk patients. The treatment initiation rate was 90% (38/42) in the group with an SFP programme versus 61% (28/46) in the control group.

Adherence

In 2016, Abbad et al.[38] published a prospective study of a type B SFP programme from a single hospital in Lille, France. Over a 4-year period, 173 eligible patients were identified, of whom 110 attended the programme. The rate of BMD testing was high at 97% (107/110), whereas the rate of treatment initiation was 35% (38/110), of whom 97% (35/37) were persistent after 21 months. Fourteen patients (13% = 14/110) sustained further fractures with a median follow-up of 20 months.

In 2011, McLellan et al.[39] published the most recent outcome data from the West Glasgow SFP programme, evaluating the rates of patient capture, assessment, treatment and cost-effectiveness over 8 years of service provision. There were 11,096 patients identified, of whom 8875 (80%) were captured or seen in the SFP programme. The number of patients who underwent BMD scans was 5405 (61%). Oral bisphosphonate treatment was recommended in 5433 patients (61%). Ninety-six percentage of patients had the recommended treatment initiated in primary care, a testimony to the effectiveness of communication with the primary care physician (PCP). Persistence with treatment was 86% at 12 months.

In 2002, Chevalley et al.[40] identified 385 patients over a 36-month period. Amongst 246 patients with complete data, 153 (62%) had a DXA scan performed, whereas 72 (29%) had osteoporosis pharmacotherapy recommended. Osteoporosis pharmacotherapy was initiated on 45 of the 72 patients (63%), whereas self-reported persistent at 6 months was 67% (30/45).

Cost-Effectiveness

In 2016, Yong et al.[41] from the Ontario Fracture Clinic Screening programme published a cost-effectiveness analysis of a type B SFP programme as well as of a type C SFP programme, which included 31 outpatient clinics. Hip fracture rates were estimated from previously published RCT data from pivotal osteoporosis pharmacotherapy trials rather than data from the service itself. The type C programme was found to be cost-effective at a cost of C$19,132 per QALY gained (2014 Canadian dollar values), based on BMD testing rate of 447 per 1000 patients (47%) versus 39 per 1000 patients (3.9%) for usual care and treatment rates of 334 per 1000 patients (33.4%) versus 106 per 1000 patients (10.6%) in the usual care group. In the type B programme, the BMD testing rate was 96%, however, the treatment rates were not reported. The type B SFP programme had greater cost-effectiveness than a type C programme at a cost of C$5720 per QALY gained. The control or usual care group for both analyses were derived from previously published Canadian data.

The cost-effectiveness analysis in 2011 by McLellan et al.[39] was based on a refracture rate of 8% at 4 years. The cost per QALY gained was £5740. According to the least favourable efficacy data, 15 fractures are avoided, at a cost of £84,076 per 1000 patients. The most favourable efficacy data demonstrated that 36 fractures are avoided and £199,132 is saved per 1000 patients.

The metaanalysis of BMD testing rates amongst these studies (Fig. 3.4) demonstrated a risk difference of 0.50 (95% CI 0.46–0.53), $P<.001$, indicating a 50% absolute difference in BMD testing rates with a type B intervention compared to usual care.

The metaanalysis of treatment initiation rates amongst these studies (Fig. 3.5) demonstrated a risk difference of 0.16 (95% CI 0.12–0.21, $P<.001$), demonstrating a 16% absolute difference in treatment initiation rates with a type B intervention compared to usual care.

Type C Model Interventions

Compared to 'type A' and 'type B' models of care, type C SFP programmes are characterised by less-intensive interventions. Once patients are identified following their minimal trauma fracture, they are made aware of their heightened risk of further osteoporotic fractures and receive education regarding other risk factors for osteoporosis in addition to lifestyle advice (including falls prevention). Although patients are informed of the need for further investigation and treatment for osteoporosis, initiation of these interventions are left to the primary care physician. Thus, the second component of type C SFP programmes involves engaging and alerting the primary care physician to the minimal trauma fracture and the need for further assessment and treatment to reduce the risk of further fractures. Communication with the patient or primary care physician can take the form of 'face-to-face' interviews, a personalised letter or a telephone call. In this model, no BMD testing or specific treatment for osteoporosis occurs within the programme. Therefore, a type C SFP programme requires fewer resources than more intensive models of care.

Outcomes

There have been 15 studies of type C interventions, of which 11 reported data with a control group relating to BMD testing or treatment initiation rates. Ten of these studies were included in the metaanalysis (Table 3.4). There were no studies that reported data relating to adherence, refractures or cost-effectiveness.

BMD Testing and Treatment Initiation Rates

In 2010 Inderjeeth et al.[42] published a 'before and after' study, from Perth, Western Australia. The intervention involved the education of primary care physicians as well as hospital physicians. After 12 months follow-up, the rate of BMD testing initiated by GPs and hospital physicians increased from 3% (6/200) in the preintervention group to 45% (39/87) in the intervention group. The initiation of specific osteoporosis pharmacotherapy increased from 6% (12/200) in the preintervention to 30% (26/87) in the intervention group. About 52% of pharmacotherapy initiation was by PCPs. Of note, the effect size was larger than expected with this type C intervention possibly because the osteoporosis education was delivered to both hospital physicians and PCPs.

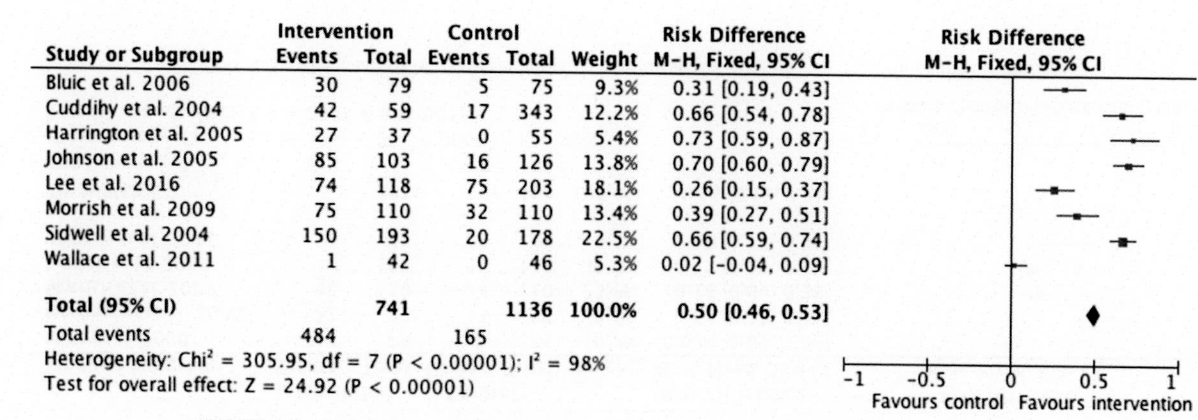

FIGURE 3.4 Metaanalysis of BMD testing rates using risk difference in intervention type B studies.

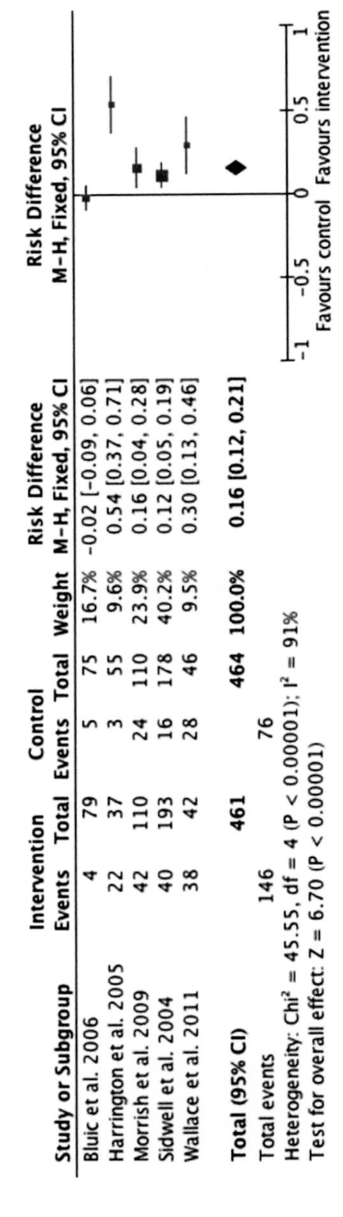

FIGURE 3.5 Metaanalysis of treatment initiation rates using risk difference in intervention type B studies.

TABLE 3.4 Summary of Studies Included in Metaanalysis of Intervention Type C Studies

Study Name	Study Type	Fracture Site	Age	% Female	BMD (Control)	BMD (Intervention)	Treatment (Control)	Treatment (Intervention)
Inderjeeth et al.[42]	Before & After	All (nil breakdown)	>65	–	6|200	18|45	12|200	14|45
Gardner et al.[43]	RCT	Hip	82 (mean)	78	6|36	12|36	6|36	10|36
Feldstein et al.[44]	RCT	All (C: hip: 8.9%, wrist 14.9%; I: hip 14.7%, wrist 15.6%)	72 (mean); >50	100	2|101	36|109	5|101	28|109
Solomon et al.[45]	RCT	All (nil breakdown)	–	–	4|95	11|134	1|95	6|134
Majumdar et al.[49]	Prospective Controlled	Wrist	66 (median)	78	8|47	34|55	5|47	22|55
Ashe et al.[48]	Prospective Controlled	Wrist	71.5 (mean)	80	5|22	11|12	–	–
Hawker et al.[47]	Before & After	All (I: wrist 64/139, hip 19/139; C: wrist 64/139, hip 25/139)	66 (mean)	74	23|139	49|139	–	–
Majumdar et al.[50]	RCT	Wrist	60 (median)	77	24|135	71|137	10|135	30|137
Cranney et al.[51]	Cluster Randomized	Wrist	69 (mean)	100	36|145	64|125	15|145	35|125
Cosman et al.[46]	Before & After	Hip	78	74	33|60	49|75	20|60	14|75

In 2005, Gardner et al.[43] reported a small prospective controlled trial of patients with hip fractures, conducted at the Hospital for Special Surgery, New York. Each group comprised 40 patients; however, 4 died during the 6 months follow-up. The 6 of 36 surviving patients in the control group had DXA scans performed, compared to 12 of 36 patients in the intervention group after 6 months. Osteoporosis pharmacotherapy was initiated in 6 of 36 patients in the control group compared to 10 of 36 in the intervention group.

In 2006, Feldstein et al.[44] reported an RCT from an HMO consisting of 454,000 members. Patients were randomised into three arms, i.e., (1) usual care, i.e., no intervention (n = 101), (2) EMR reminder to the primary care physician (n = 101) and (3) EMR reminder to the primary care physician with patient education, i.e., a type C model intervention (n = 109). After 6 months of follow-up, the rate of BMD testing was 2% (2/101) in the control group and 33% (36/109) in the type C intervention. The rate of osteoporosis pharmacotherapy initiation was 5% (5/101) in the control group and 20% (22/109) with the type C intervention. There was no difference in results between the two intervention groups.

In 2007, Solomon et al.[45] performed an RCT of a type C intervention versus usual care (control group) within a large health-care insurer, Horizon Blue Cross Blue Shield of New Jersey. After 10 months of follow-up, there was a small, statistically significant but not clinically significant improvement in BMD testing from 9% (86/976) in the control group to 13% (126/997) in the intervention group. Similarly, initiation of pharmacotherapy was slightly higher in the intervention group, at 6% (59/997) versus 4% (36/976) in the control group. There were no differences in the incidence of refractures; however, the actual numbers were not reported.

In 2017, Cosman et al.[46] published outcomes before and after the institution of a type C intervention for acute hip fracture patients at Helen Hayes Hospital, New York, USA. In the preintervention group, the rate of BMD testing was 55% (33/60) compared to 65% (49/75) after the intervention. In the preintervention group, osteoporosis treatment rates were 38% prefracture and 33% after the hip fracture. In the postintervention group, osteoporosis treatment rates were 21% before and 19% after the hip fracture. That is, in this study the rate of osteoporosis pharmacotherapy remained poor with the type C intervention, indicating a need for a more intensive SFP programme.

In 2003, Hawker et al.[47] published the first study of a type C SFP programme from Canada. The setting was in five community hospitals in the Toronto area. It only included treatment naïve patients. The preintervention group was age and sex matched to the intervention group (n = 139 in each group). The rate of BMD testing was 17% (23/139) preintervention and 35% (49/139) in the intervention group. The rate of osteoporosis treatment initiation was similar in both groups i.e., 10% (14/139) in the control versus 11% (15/139) in the intervention group.

A small Vancouver-based prospective controlled study published in 2004 by Ashe et al.[48] included patients with incident wrist fractures. The rate of BMD testing in the control arm was 5/22 versus 11/12 in the intervention arm. Treatment initiation rates were not reported.

In 2004, Majumdar et al.[49] described a type C intervention amongst treatment naïve patients with incident wrist fractures in a small nonrandomised controlled trial conducted in Edmonton, Alberta. Six months after the fracture, the rate of BMD testing was 17% (8/47) in the control group versus 62% (34/55) in the intervention group. Osteoporosis treatment rates were higher in the intervention group compared to the control group at 40% (22/55) versus 11% (5/47), respectively.

In 2008, the same group[50] published an RCT of treatment naive patients with incident wrist fractures. Six months after the fracture, intervention patients were more likely to undergo BMD testing (52% or 71/137 vs. 18% or 24/135) and treatment initiation (22% or 30/137 vs. 7% or 10/135).

In 2008, Cranney et al.[51] reported a cluster-randomised trial which included patients with wrist fractures in the Ontario area. After 6 months, the rate of BMD testing was 53% (64/120) versus 26% (36/141) in the intervention versus control groups, respectively. Similarly, the intervention was associated with a higher rate of treatment initiation, i.e., 29% (35/120) in the intervention group versus 11% (15/141) in the control group.

The metaanalysis of BMD testing rates amongst these model C studies (Fig. 3.6) demonstrated a risk difference of 0.25 (95% CI 0.21–0.29, $P<.001$). In other words, there was a 25% absolute difference in BMD testing rates with a type C intervention compared to usual care.

The metaanalysis of treatment initiation rates amongst these studies (Fig. 3.7), demonstrated a risk difference of 0.13 (95% CI 0.09–0.16, $P<.001$), indicating a 13% absolute difference in treatment initiation rates with a type C intervention compared to usual care.

Type D Model Interventions

'Type D' SFP programmes represent interventions whereby patients receive specific education relating to osteoporosis following a minimal trauma fracture. This can be in the form of a letter, educational pamphlet, or direct communication with the patient via a 'face-to-face' interaction or a telephone call. Of note, the primary care physician is not involved in this intervention.

Two studies from Australia describe the type D intervention (Table 3.5). A randomized controlled trial in 2006 by Bliuc et al.[30] has been described in detail above under 'Type B SFP programmes'. This trial compared a type B intervention with a type D intervention amongst 80 and 79 patients respectively. The type B intervention involved a letter with an offer for a free BMD test, whereas the type D intervention meant that the patients received a letter only. The rate of BMD testing for type B versus D interventions was 38% (30/79) versus 7% (5/75), respectively, whereas the rate of treatment initiation was similar at 5% (4/79) versus 7% (5/75), respectively.

A cross-sectional study in 2002 by Diamond and Lindenburg[52] identified patients from a private radiology practice who had sustained a minimal trauma

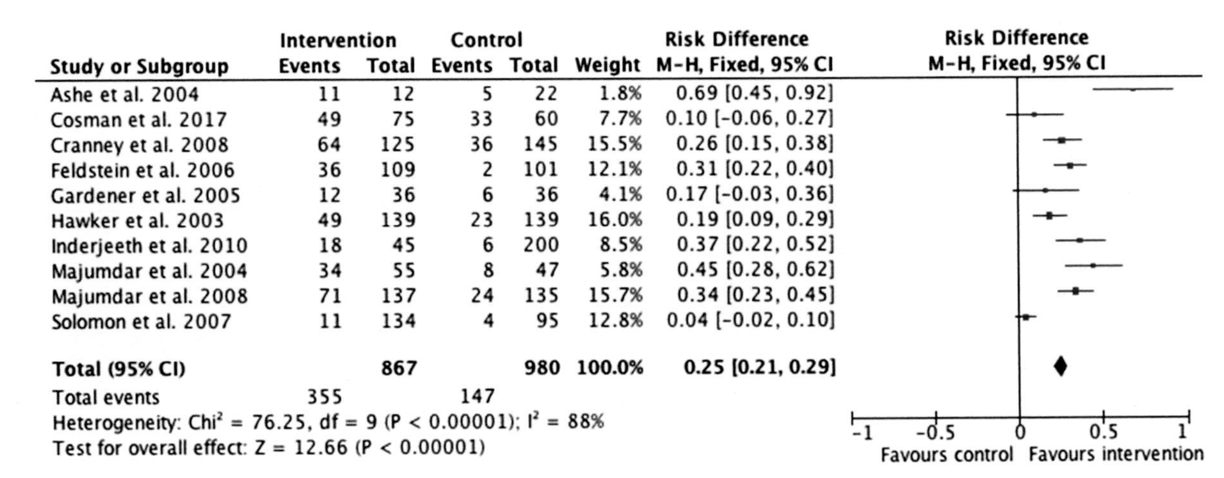

FIGURE 3.6 Metaanalysis of BMD testing rates using risk difference in intervention type C studies.

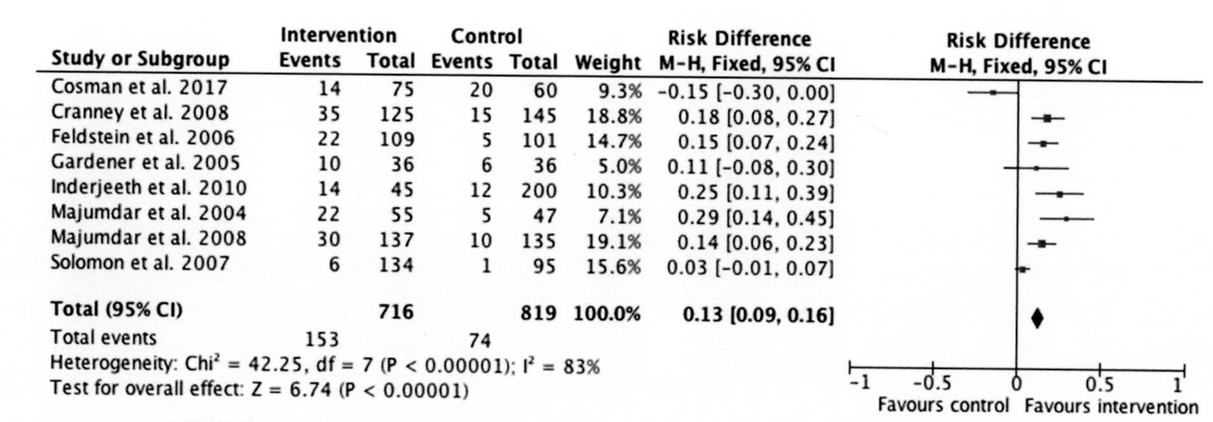

FIGURE 3.7 Metaanalysis of treatment initiation rates using risk difference in intervention type C studies.

TABLE 3.5 Summary of Intervention Type D Studies

Country	Study Name	Study Type	Settings	Identification Methods	Fracture Site	Age	Female (%)	N	BMD (Control)	BMD (Intervention)	Treatment (Control)	Treatment (Intervention)
Australia	Diamond T and Lindenburg M[52]	Cross-sectional	Radiology	Radiology records	All	76 (mean)	64	161	–	82|161	–	46|161
Canada	Bessette et al.[54]	RCT	OP	EMR (database)	All	62 (mean)	100	1174	–	–	31|386	90|788

fracture. These patients were asked to fill out a questionnaire and given an information card that encouraged them to speak to their PCP regarding BMD testing and antiosteoporosis therapy. About 64% (161/250) of eligible patients were contacted. The rate of BMD testing was 51% (82/161) at 12 months, whilst 29% (46/161) of patients were receiving specific osteoporosis pharmacotherapy. The larger than expected effect size with this type D intervention may have been due to selection bias. Those patients who filled out the questionnaire may have been more likely to talk to their PCP's regarding their bone health.

SUMMARY AND CONCLUSIONS

This chapter attempts to compare the outcomes of different types of SFP programmes classified according to their intervention intensity, i.e., from the most intensive model (type A) to least intensive model of care (type D). The outcomes from the metaanalyses are summarised in Table 3.6, Figs 3.8 and 3.9. Comparisons between models are generally difficult because of a lack of standardised outcome measures. Nevertheless, Table 3.6 demonstrates a clear relationship between intervention intensity and intervention outcomes in terms of BMD testing and treatment initiation rates. As common sense would suggest, more intensive interventions produce better outcomes.

We reported similar results in 2012 (Ganda et al.)[1]; however, the current, more up-to-date metaanalysis not only confirms but also strengthens our previous findings, as demonstrated by narrower confidence intervals in each of the metaanalyses. Unfortunately, there is a paucity of data relating to type D interventions and therefore the associated studies could not be included in the metaanalysis.

It is important to note that BMD testing and treatment initiation rates are only process measures. The clinically relevant outcome measure is refractures.

TABLE 3.6 Summary of Metaanalyses

	BMD Testing			Treatment Initiation		
Intervention Type	No. of Studies	Risk Difference (95% CI)	*P*	No. of Studies	Risk Difference (95% CI)	*P*
Type 'A'	8	0.58 (0.54–0.62)	<.001	13	0.29 (0.26–0.32)	<.001
Type 'B'	8	0.50 (0.46–0.53)	<.001	5	0.16 (0.12–0.21)	<.001
Type 'C'	10	0.25 (0.21–0.29)	<.001	8	0.13 (0.09–0.16)	<.001

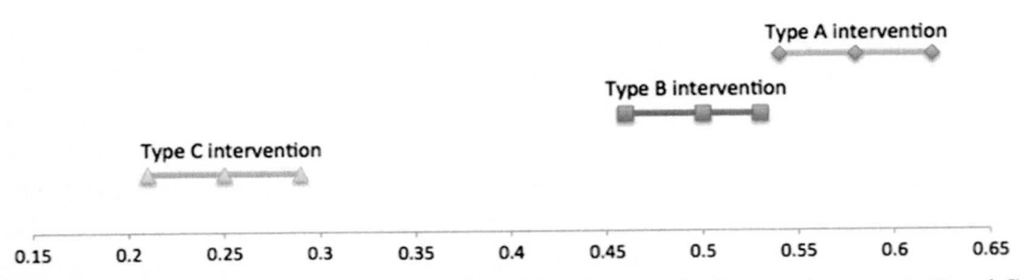

FIGURE 3.8 Summary of metaanalyses of BMD testing rates for intervention type A, B and C studies; x-axis: risk difference.

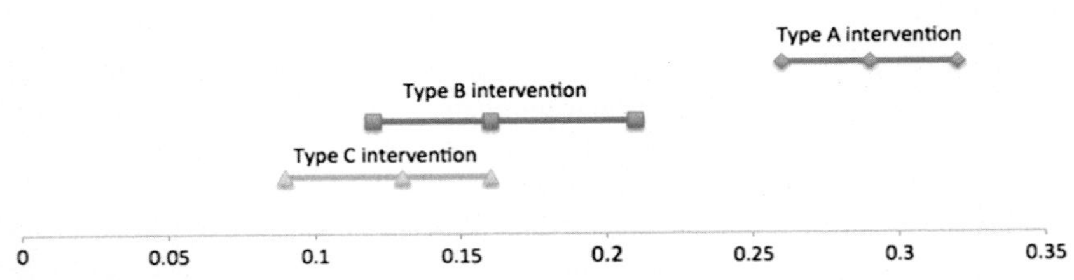

FIGURE 3.9 Summary of metaanalyses of treatment initiation rates for intervention type A, B and C studies; x-axis: risk difference.

However, data on refracture rates are scarce and generally limited to type A interventions.

Despite these findings, a type A model of care may not be suitable for all health care settings because of limitations in resources. Thus, the specific health care system in which an SFP program is embedded is of pivotal importance. For example, a type B intervention has been shown to be extremely effective in the United Kingdom because of the strong structural integration between primary care physicians and public hospitals. As would be expected, educational interventions alone (type C or D models) were less effective than type A or B programs. An educational intervention system still had some limited benefits and may therefore be an option in poor resource areas ('Better than nothing').

A major deficit in the published literature on models of postfracture care is the inconsistent reporting of results. This covers a spectrum of outcomes such as the identification rate of potentially eligible patients, the length of time between fracture and SFP program evaluation, the extent of risk factor evaluation and the assessment for secondary causes for osteoporosis. Moreover, the rate of treatment initiation may be reported including or excluding treatment naïve patients. Adherence to antiosteoporosis therapy may be self-reported or based on pharmaceutical claims data. Therefore, it is difficult to compare between studies, a fact reflected by the relatively high degree of heterogeneity in the metaanalyses. Standardised reporting of these outcome measures would be important for quality assurance and to benchmark performance, as well as in the comparison of different studies.

The interpretation of the outcomes from the literature review and metaanalysis should therefore be viewed in the context of significant heterogeneity in the reporting of outcome measures between studies. Commonly reported outcome measures described are BMD testing rates, treatment initiation rates, adherence, refracture rates and cost-effectiveness analyses. Thus, treatment initiation rates are reported in varying ways with some studies using different denominators (captured patients vs. identified patients) or including treatment naïve subjects versus those already on treatment prior to the fracture. Furthermore, there was heterogeneity in study types, which ranged from retrospective audits and prospective studies, to studies with control groups, which were either historical, concurrent (nonattenders) or a service/hospital of similar size without an SFP program.

Treatment rates may vary between health networks and countries because of different diagnostic and treatment thresholds. For example, in the South Glasgow service, patients with hip fractures were treated without a prior BMD scan, whereas in the West Glasgow service a BMD study was routinely performed in all patients and pharmacotherapy was initiated if the T-score was less than −2.0 SD. This threshold, however, differs from that of other centres (e.g., Majumdar et al.[5]: T-score of less than −1.5 SD). In the UK studies, only oral bisphosphonates were initiated, mostly because of cost considerations. Those patients who could not tolerate or receive oral bisphosphonates were treated with calcium and vitamin D only, a strategy, which would certainly differ between countries.

Osteoporosis care is still suboptimal, even in places where 'type A' services are available. Thus, most SFP programs only capture patients with peripheral (nonvertebral) fractures, whereas vertebral fractures are often asymptomatic and therefore harder to identify. To improve capture rates, SFP programs will need to utilize integrated electronic health system databases and intelligent data mining tools.

Also, attempts should be made at collaborations between centres, especially in fragmented health-care networks within countries. Although SFP programs have contributed significantly towards closing the care gap in osteoporosis management, there is still room for significant improvement in postfracture management of patients with osteoporosis.

REFERENCES

1. Ganda K, Puech M, Chen JS, Speerin R, Bleasel J, Center JR, et al. Models of care for the secondary prevention of osteoporotic fractures: a systematic review and meta-analysis. *Osteoporos Int* 2013;**24**(2):393–406.
2. Vaile J, Sullivan L, Bennett C, Bleasel J. First fracture project: addressing the osteoporosis care gap. *Intern Med J* 2007;**37**(10):717–20.
3. Kuo I, Ong C, Simmons L, Bliuc D, Eisman J, Center J. Successful direct intervention for osteoporosis in patients with minimal trauma fractures. *Osteoporos Int* 2007;**18**(12):1633–9.
4. Yates CJ, Chauchard MA, Liew D, Bucknill A, Wark JD. Bridging the osteoporosis treatment gap: performance and cost-effectiveness of a fracture liaison service. *J Clin Densitom* 2015;**18**(2):150–6.

5. Majumdar SR, Beaupre LA, Harley CH, Hanley DA, Lier DA, Juby AG, et al. Use of a case manager to improve osteoporosis treatment after hip fracture. *Arch Intern Med* 2007;**167**(19):2110–5.
6. Majumdar SR, Johnson JA, Bellerose D, McAlister FA, Russell AS, Hanley DA, et al. Nurse case-manager vs multifaceted intervention to improve quality of osteoporosis care after wrist fracture: randomized controlled pilot study. *Osteoporos Int* 2011;**22**(1):223–30.
7. Streeten EA, Mohamed A, Gandhi A, Orwig D, Sack P, Sterling R, et al. The inpatient consultation approach to osteoporosis treatment in patients with a fragility fracture: is automatic consultation needed? *J Bone Joint Surg Am* 2006;**88-A**(9):1968–74.
8. Queally JM, Kiernan C, Shaikh M, Rowan F, Bennett D. Initiation of osteoporosis assessment in the fracture clinic results in improved osteoporosis management: a randomised controlled trial. *Osteoporos Int* 2013;**24**(3):1089–94.
9. Ruggiero C, Zampi E, Rinonapoli G, Baroni M, Serra R, Zengarini E, et al. Fracture prevention service to bridge the osteoporosis care gap. *Clin Interv Aging* 2015;**10**:1035–42.
10. Lih A, Nandapalan H, Kim M, Yap C, Lee P, Ganda K, et al. Targeted intervention reduces refracture rates in patients with incident non-vertebral osteoporotic fractures: a 4-year prospective controlled study. *Osteoporos Int* 2011;**22**(3):849–58.
11. Van der Kallen J, Giles M, Cooper K, Gill K, Parker V, Tembo A, et al. A fracture prevention service reduces further fractures two years after incident minimal trauma fracture. *Int J Rheum Dis* 2014;**17**(2):195–203.
12. Edwards BJ, Bunta AD, Madison LD, DeSantis A, Ramsey-Goldman R, Taft L, et al. An osteoporosis and fracture intervention program increases the diagnosis and treatment for osteoporosis for patients with minimal trauma fractures. *Jt Comm J Qual Patient Saf* 2005;**31**(5):267–74.
13. Olenginski TP, Maloney-Saxon G, Matzko CK, Mackiewicz K, Kirchner HL, Bengier A, et al. High-risk osteoporosis clinic (HiROC): improving osteoporosis and postfracture care with an organized, programmatic approach. *Osteoporos Int* 2015;**26**(2):801–10.
14. Ganda K, Schaffer A, Pearson S, Seibel MJ. Compliance and persistence to oral bisphosphonate therapy following initiation within a secondary fracture prevention program: a randomised controlled trial of specialist vs. non-specialist management. *Osteoporos Int* 2014;**25**(4):1345–55.
15. Dehamchia-Rehailia N, Ursu D, Henry-Desailly I, Fardellone P, Paccou J. Secondary prevention of osteoporotic fractures: evaluation of the Amiens University Hospital's fracture liaison service between January 2010 and December 2011. *Osteoporos Int* 2014;**25**(10):2409–16.
16. Naranjo A, Ojeda-Bruno S, Bilbao-Cantarero A, Quevedo-Abeledo JC, Diaz-Gonzalez BV, Rodriguez-Lozano C. Two-year adherence to treatment and associated factors in a fracture liaison service in Spain. *Osteoporos Int* 2015;**26**(11):2579–85.
17. Goltz L, Degenhardt G, Maywald U, Kirch W, Schindler C. Evaluation of a program of integrated care to reduce recurrent osteoporotic fractures. *Pharmacoepidemiol Drug Saf* 2013;**22**(3):263–70.
18. Boudou L, Gerbay B, Chopin F, Ollagnier E, Collet P, Thomas T. Management of osteoporosis in fracture liaison service associated with long-term adherence to treatment. *Osteoporos Int* 2011;**22**(7):2099–106.
19. Eekman DA, Helden SH, Huisman AM, Verhaar HJJ, Bultink IEM, Geusens PP, et al. Optimizing fracture prevention: the fracture liaison service, an observational study. *Osteoporos Int* 2014;**25**(2):701–9.
20. Ganda K, Schaffer A, Seibel MJ. Predictors of re-fracture amongst patients managed within a secondary fracture prevention program: a 7-year prospective study. *Osteoporos Int* 2015;**26**(2):543–51.
21. Nakayama A, Major G, Holliday E, Attia J, Bogduk N. Evidence of effectiveness of a fracture liaison service to reduce the re-fracture rate. *Osteoporos Int* 2016;**27**(3):873–9.

22. Huntjens KM, van Geel TA, van den Bergh JP, van Helden S, Willems P, Winkens B, et al. Fracture liaison service: impact on subsequent nonvertebral fracture incidence and mortality. *J Bone Joint Surg Am* 2014;**96**(4):e29.
23. Lyles KW, Colon-Emeric CS, Magaziner JS, Adachi JD, Pieper CF, Mautalen C, et al. Zoledronic acid and clinical fractures and mortality after hip fracture. *N Engl J Med* 2007;**357**(18):1799–809.
24. Dell R, Greene D, Schelkun SR, Williams K. Osteoporosis disease management: the role of the orthopaedic surgeon. *J Bone Joint Surg Am* 2008;**90**(Suppl. 4):188–94.
25. Cooper MS, Palmer AJ, Seibel MJ. Cost-effectiveness of the Concord Minimal Trauma Fracture Liaison service, a prospective, controlled fracture prevention study. *Osteoporos Int* 2012;**23**(1):97–107.
26. Sander B, Elliot-Gibson V, Beaton DE, Bogoch ER, Maetzel A. A coordinator program in post-fracture osteoporosis management improves outcomes and saves costs. *J Bone Joint Surg Am* 2008;**90**(6):1197–205.
27. Bogoch E, Elliot-Gibson V, Beaton D, Jamal S, Josse R, Murray T. Effective initiation of osteoporosis diagnosis and treatment for patients with a fragility fracture in an orthopaedic environment. *J Bone Joint Surg Am* 2006;**88-A**(1):25–34.
28. Majumdar SR, Lier DA, Beaupre LA, Hanley DA, Maksymowych WP, Juby AG, et al. Osteoporosis case manager for patients with hip fractures: results of a cost-effectiveness analysis conducted alongside a randomized trial. *Arch Intern Med* 2009;**169**(1):25–31.
29. Mitchell PJ. Fracture liaison services in the United Kingdom. *Curr Osteoporos Rep* 2013;**11**(4):377–84.
30. Bliuc D, Eisman JA, Center JR. A randomized study of two different information-based interventions on the management of osteoporosis in minimal and moderate trauma fractures. *Osteoporos Int* 2006;**17**(9):1309–17.
31. Cuddihy MT, Amadio PC, Gabriel SE, Pankratz VS, Kurland RL, Melton 3rd LJ. A prospective clinical practice intervention to improve osteoporosis management following distal forearm fracture. *Osteoporos Int* 2004;**15**(9):695–700.
32. Harrington JT, Barash HL, Day S, Lease J. Redesigning the care of fragility fracture patients to improve osteoporosis management: a health care improvement project. *Arthritis Rheum* 2005;**53**(2):198–204.
33. Johnson SL, Petkov VI, Williams MI, Via PS, Adler RA. Improving osteoporosis management in patients with fractures. *Osteoporos Int* 2005;**16**(9):1079–85.
34. Lee RH, Pearson M, Lyles KW, Jenkins PW, Colon-Emeric C. Geographic scope and accessibility of a centralized, electronic consult program for patients with recent fracture. *Rural Remote Health* 2016;**16**(1):3440.
35. Morrish DW, Beaupre LA, Bell NR, Cinats JG, Hanley DA, Harley CH, et al. Facilitated bone mineral density testing versus hospital-based case management to improve osteoporosis treatment for hip fracture patients: additional results from a randomized trial. *Arthritis Rheum* 2009;**61**(2):209–15.
36. Sidwell AI, Wilkinson TJ, Hanger HC. Secondary prevention of fractures in older people: evaluation of a protocol for the investigation and treatment of osteoporosis. *Intern Med J* 2004;**34**:129–32.
37. Wallace I, Callachand F, Elliott J, Gardiner P. An evaluation of an enhanced fracture liaison service as the optimal model for secondary prevention of osteoporosis. *JRSM Short Rep* 2011;**2**(2):8.
38. Abbad N, Lemeunier L, Chantelot C, Puisieux F, Cortet B. Secondary prevention program for osteoporotic fractures at Lille University Hospital. *Presse Med* 2016;**45**(3):375–7.

39. McLellan AR, Wolowacz SE, Zimovetz EA, Beard SM, Lock S, McCrink L, et al. Fracture liaison services for the evaluation and management of patients with osteoporotic fracture: a cost-effectiveness evaluation based on data collected over 8 years of service provision. *Osteoporos Int* 2011;**22**:2083–98.
40. Chevalley T, Hoffmeyer P, Bonjour JP, Rizzoli R. An osteoporosis clinical pathway for the medical management of patients with low-trauma fracture. *Osteoporos Int* 2002;**13**:450–5.
41. Yong JH, Masucci L, Hoch JS, Sujic R, Beaton D. Cost-effectiveness of a fracture liaison service–a real-world evaluation after 6 years of service provision. *Osteoporos Int* 2016;**27**(1):231–40.
42. Inderjeeth CA, Glennon DA, Poland KE, Ingram KV, Prince RL, Van VR, et al. A multimodal intervention to improve fragility fracture management in patients presenting to emergency departments. *Med J Aust* 2010;**193**(3):149–53.
43. Gardner MJ, Brophy RH, Demetrakopoulos D, Koob J, Hong R, Rana A, et al. Interventions to improve osteoporosis treatment following hip fracture. *J Bone Joint Surg Am* 2005;**87-A**(1):3–7.
44. Feldstein A, Elmer PJ, Smith DH, Herson M, Orwoll E, Chen C, et al. Electronic medical record reminder improves osteoporosis management after a fracture: a randomized, controlled trial. *J Am Geriatr Soc* 2006;**54**(3):450–7.
45. Solomon DH, Polinski JM, Stedman M, Truppo C, Breiner L, Egan C, et al. Improving care of patients at-risk for osteoporosis: a randomized controlled trial. *J Gen Intern Med* 2007;**22**(3):362–7.
46. Cosman F, Nicpon K, Nieves JW. Results of a fracture liaison service on hip fracture patients in an open healthcare system. *Aging Clin Exp Res* 2017;**29**(2):331–4.
47. Hawker G, Ridout R, Ricupero M, Jaglal S, Bogoch E. The impact of a simple fracture clinic intervention in improving the diagnosis and treatment of osteoporosis in fragility fracture patients. *Osteoporos Int* 2003;**14**(2):171–8.
48. Ashe M, Khan K, Guy P, Kruse K, Hughes K, O'Brien P, et al. Wristwatch-distal radial fracture as a marker for osteoporosis investigation: a controlled trial of patient education and a physician alerting system. *J Hand Ther* 2004;**17**(3):324–8.
49. Majumdar SR, Rowe BH, Johnson JA, Holroyd BH, Morrish DW, Maksymowych WP, et al. A controlled trial to increase detection and treatment of osteoporosis in older patients with a wrist fracture. *Ann Intern Med* 2004;**141**(5):366–73.
50. Majumdar SR, Johnson JA, McAlister FA, Bellerose D, Russell AS, Hanley DA, et al. Multifaceted intervention to improve diagnosis and treatment of osteoporosis in patients with recent wrist fracture: a randomized controlled trial. *CMAJ* 2008;**178**(5):569–75.
51. Cranney A, Lam M, Ruhland L, Brison R, Godwin M, Harrison MM, et al. A multifaceted intervention to improve treatment of osteoporosis in postmenopausal women with wrist fractures: a cluster randomized trial. *Osteoporos Int* 2008;**19**(12):1733–40.
52. Diamond T, Lindenburg M. Osteoporosis detection in the community. *Aust Fam Physician* 2002;**31**(8):751–752.
53. Jones G, Warr S, Francis E, Greenaway T. The effect of a fracture protocol on hospital prescriptions after minimal trauma fractured neck of the femur: a retrospective audit. *Osteoporos Int* 2005;**16**(10):1277–80.
54. Bessette L, Davison KS, Jean S, Roy S, Ste-Marie LG, Brown JP. The impact of two educational interventions on osteoporosis diagnosis and treatment after fragility fracture: a population-based randomized controlled trial. *Osteoporos Int* 2011;**22**(12):2963–72.

Chapter 4

Fracture Liaison Services: An Australasian Perspective

Jacqueline C.T. Close
Neuroscience Research Australia and Prince of Wales Hospital Clinical School, University of New South Wales, Sydney, Australia

INTRODUCTION

This chapter puts a spotlight on fracture liaison services in Australasia. Australasia is a region of Oceania, comprising Australia, New Zealand and the island of New Guinea. For the purposes of this chapter, we will focus on the countries of Australia and New Zealand.

The evidence base for secondary fracture prevention and fracture liaison services is robust. The challenge across the world is to use this knowledge and ensure that people who sustain a low-trauma fracture get access to treatments from which they stand to benefit. Addressing and closing the 'care gap' requires skills, knowledge and expertise beyond the clinical interventions known to alter fracture risk. The challenge is how to make it happen in a systematic manner that ensures people at risk are identified, investigated and offered treatments appropriate to their needs. Solutions require an understanding of the country, its politics and particularly its health system. It needs an overarching strategy and an appropriately resourced high-level implementation plan led by people with the skill set to navigate the complexities and nuances of health systems and who can opportunistically push forward the secondary fracture prevention agenda. Failure to achieve this leads to fragmented care with islands of high-quality care in a sea of mediocrity.

Australia

Australia has a population of over 24 million and a population density of 3.2 per km^2.[1,2] Whilst most of the population is centred around coastal towns and cities, a sizeable percentage of the population lives in rural and remote settings. Colonised by the United Kingdom in 1788, Australia has evolved from six self-governing areas to a Federation of Australia comprising six States and two Territories. Power to govern is shared between State/Territory

Secondary Fracture Prevention. https://doi.org/10.1016/B978-0-12-813136-7.00004-1

and Commonwealth Governments with parliamentary processes based on a Westminster system. Health care is largely provided by private health professionals or by private or State government-funded hospitals. Medicare is Australia's universal health-care system, which is the primary health scheme that subsidises medical costs in Australia. In addition to Medicare, there is the Pharmaceutical Benefits Scheme funded by the Commonwealth government which subsidises a range of prescription medications including most treatments for osteoporosis.

Figures from the Australian Institute of Health and Welfare provide estimates of the extent of osteoporosis in Australia.[3,4] Approximately 1 in 10 Australians aged 50 years and older have osteoporosis or osteopenia, and in 2013–14 there were 62,132 hospitalisations for low-trauma fracture.

In 2013, Osteoporosis Australia published a report providing information on the current and future human and financial costs of osteoporosis from 2012–22.[5] The data provide a stark reminder of the impact of an ageing population and our failure to address and treat osteoporosis including secondary fracture prevention. The number of fragility fractures at all anatomical sites is projected to increase from 144,312 in 2013 to 183,105 in 2022. Despite a levelling off of the rate of hip fracture, the total number of hip fractures over the same period is estimated to increase from 23,990 to 32,413 with a predicted cost for hip fracture alone of over $1 billion.[5]

New Zealand

New Zealand is an island country situated approximately 1500 km east of Australia with a population of ~4.7 million in 2016.[6] Like Australia, New Zealand parliamentary processes are based on a Westminster system. However, unlike the confederated nature of Australia, it has a single government.[7] The health-care system of New Zealand is a largely publicly funded system, although in more recent years private practice has become more commonplace. Hospital and specialist care in New Zealand is covered by the government if the patient is referred by a general or family practitioner. The Ministry of Health is responsible for the oversight and funding of all the District Health Boards (DHBs). The Pharmaceutical Management Agency of New Zealand (PHARMAC) was set up in 1993 to decide which medications the government will subsidise.[8] In general, PHARMAC will select an effective and safe medication from a class of drugs and negotiate with the drug manufacturer to obtain the best price. There are approximately 2000 drugs listed on the national schedule that are either fully or partially subsidised.[8]

There were an estimated 84,354 osteoporotic fractures in New Zealand in 2007, including 3803 hip and 27,994 vertebral fractures with an estimated direct cost of $NZ330 million. Those figures are projected to rise by over 30% to 115,914 fractures in 2020 with an associated cost of $NZ458 million and a loss of 15,176 quality adjusted life years.[9]

THE OSTEOPOROSIS CARE GAP

As with many countries across the world, there is no shortage of information from Australia and New Zealand to tell us that when it comes to secondary fracture prevention, we are not doing well. There is an abundance of literature in various formats that highlights the 'care gap' and what impact that care gap has on the population. Studies published at a State or Territory level in Australia including New South Wales,[10–14] South Australia,[15] Australian Capital Territory,[16] Victoria[17,18] and Western Australia[19,20] all give a fairly consistent message around failure to effectively identify and manage osteoporosis and secondary fracture prevention. New Zealand paints a similar picture of missed opportunity.[21–23]

Published in 2004, The Australian BoneCare study surveyed over 88,000 postmenopausal women from 927 general practitioners with a view to exploring the recognition and treatment of osteoporosis.[24] Of the study sample, 29% had sustained a prior low-trauma fracture, and the majority of these women (72%) were not on treatment for osteoporosis.

In 2009, Chen et al. published a paper looking at the management of osteoporosis in primary care and interviewed 37,957 people with or at risk of developing chronic disease. They found that of those who had already sustained a low-trauma fracture, 29.7% were receiving treatment for osteoporosis. Figures for treatment dropped to 3.8% when looking at people with radiographic evidence of prior vertebral fracture, highlighting the missed opportunity of identifying at-risk populations earlier in the disease process.[25]

A number of high-level factors also appear to impact on the care gap. In a study of hip fracture, patients in Western Australia using residential postcodes, data showed that regional and rural residents were 65% less likely to use vitamin D, 60% less likely to use calcium and 46% less likely to use a bisphosphonate before hip fracture.[26] We also know that people living in residential aged care facilities are at high risk of falls and fracture but are often less likely to be offered treatment to prevent fractures. This has been acknowledged through a consensus paper in Australia providing recommendations for fracture prevention for residential aged care facilities.[27]

Australia and New Zealand have indigenous peoples with the Aboriginal population comprising the large majority if indigenous people in Australia. A paper by Wong et al. identified a widening gap in the rate of hip fracture between indigenous and nonindigenous people in Western Australia with a 7.2% annual increase in hip fracture rate over the study period in indigenous people versus a decrease of 3.4% per year in nonindigenous people.[28]

CLOSING THE OSTEOPOROSIS CARE GAP

It is clear that the presentation of facts alone is insufficient to effect change. Achieving meaningful change at a whole of system level requires a multipronged approach, bringing together people and organisations with different skills, knowledge and expertise. Knowledge of the clinical issue is important, but understanding the barriers and enablers to system-level change is a key to producing generalisable and sustainable solutions.

In recent years, Australia and New Zealand are seeing a more concerted and collaborative approach to secondary fracture prevention including trans-Tasman publications. The Australian and New Zealand Bone and Mineral Society released a Position Paper and Call to Action in 2015 advocating for rapid implementation of secondary fracture prevention services.[29] The document was endorsed by a large number of organisations all with the shared goal of reducing the human cost of osteoporosis in Australia and New Zealand.

New Zealand

BoneCare 2020

Osteoporosis New Zealand (osteoporosis.org.nz) is a registered charitable trust with a clear vision – '*better bones and fewer fractures for New Zealanders*'. Its stated mission is '*To engage with the public, health professionals, policymakers and the private sector, through programmes of awareness, advocacy and education, to prevent fractures caused by osteoporosis*'. In December 2012, Osteoporosis New Zealand published BoneCare 2020.[23] It made a strong case for a systematic approach to hip fracture care and prevention to be implemented nationally. It highlighted the missed opportunities in our systems of care and reminds us that approximately 50% of people with a hip fracture have had a previous low-trauma fracture and have essentially told us they are coming. With a goal of moving from a starting point of usual care being no care, the document pushes to ensure that '*every patient presenting to urgent care services in New Zealand with a fragility fracture receives appropriate osteoporosis management and falls assessment to reduce their future fracture risk*'.

Living Stronger for Longer

In 2016, the New Zealand Accident Compensation Corporation, Ministry of Health and the Health Quality and Safety Commission agreed to work towards a national integrated approach to preventing falls and fractures, supported by an outcomes framework designed to drive and measure change.[30] This includes an agreed national minimum data set of indicators which will expand over time. The initial set of indicators include fall-related injury (fracture and nonfracture), fall-related hospitalisations (fracture and nonfracture), in-patient falls, time to surgery for hip fracture patients and uptake of vitamin D in residential aged care facilities. The approach is ambitious and, if successful, provides strong evidence for a whole of system approach to addressing falls and fracture prevention.

Better Bones, Fewer Fractures

In 2017, Osteoporosis New Zealand released its guidance for fracture prevention and management – 'Better Bones, fewer fractures'. It offers simple, practical advice on the management of osteoporosis including a clinical algorithm to assist with treatment decisions.[31]

A key factor in New Zealand's strategy is the collaborative approach that has emerged through acknowledging that falls prevention, secondary fracture prevention and hip fracture care are all pieces of the same jigsaw. The approach has been both opportunistic and pragmatic with acknowledgement of a lot of the good work that has occurred internationally and a focus on the approach to implementation rather than reinventing wheels. It is likely that New Zealand will be seen in the future as a microcosm of success in relation to secondary fracture prevention. Over a relatively short period of time, it has put in place the necessary building blocks to ensure that there is and will be a nationally consistent approach to secondary fracture prevention. At the time of writing this chapter, 16 of 20 DHBs had set up fracture liaison services – a remarkable achievement over a short time frame. As the number of fracture liaison services continues to increase, we await the data to show that the approach is making a difference.

Australia

National Health Priority Area

In 2002, the Australian Health Ministers identified musculoskeletal conditions including osteoporosis as a National Health Priority Area.[32] Subsequently in 2005, a National Action Plan for Osteoarthritis, Rheumatoid Arthritis and Osteoporosis was released with a recommendation around promotion of post-fracture assessment to reduce future fracture risk whilst the associated National Service Improvement Framework highlighted the stark care gap evident at the time.[33,34] However, despite this, there has been little change at a national level with osteoporosis remaining underdiagnosed and undertreated. In 2016, a number of key professional organisations involved in the prevention and management of osteoporosis and fractures released a National Action Plan with a focus on (1) increasing awareness and support, (2) improving osteoporosis prevention and treatment and (3) finding a cure for osteoporosis. Twenty recommendations were made including a specific recommendation to 'drive the implementation and funding of fracture liaison services in the hospital and primary care setting to prevent further fractures'. However, whilst making sound recommendations, there is little contained within the document as to who or what organisation is responsible for owning the issue and driving change.[35] An additional challenge seen in Australia with regard to ownership, responsibility and accountability is the fragmented approach to the funding and delivery of health care. Management of chronic disease in the community is largely a Commonwealth responsibility, yet the acute care costs of osteoporosis with respect to fracture are largely borne by hospitals and therefore the State and Territory Health departments. This was recently acknowledged in a national report by the Primary Health Care Advisory Group to the Australian Government on *Better Outcomes for People with Chronic and Complex Health Conditions.*[36] It states that 'Our current health system is not optimally set up to effectively manage long-term conditions'. The report also reveals that patients often experience

- a fragmented system, with providers and services working in isolation from each other rather than as a team;
- uncoordinated care;
- at times, service duplication; at other times, absent or delayed services;
- a low uptake of digital health and other health technology by providers to overcome these barriers;

SOS Fracture Alliance – A National Approach to Secondary Fracture Prevention

The Australian SOS Fracture Alliance was launched in June 2017 and brings together 32 medical, allied health, patient advocacy, carer and other organisations under its umbrella.[37] With more than 2.91 million individual members, it has a simple goal – 'to make the first break the last' by improving the care of patients presenting with an osteoporotic fracture. Borne out of years of frustration from multiple organisations trying with varying degrees of success to address the same care gap, the Alliance represents the coming together of like-minded people and organisations with the desire to work collaboratively to address this public health issue across the nation. The work of the Alliance is in its infancy in Australia, and it is too early to make predictions on the likelihood of success or otherwise of this collaboration.

Islands of Good Practice in a Sea of Mediocrity

There are undoubtedly pockets of good practice in Australia where fracture liaison services have been set up and offer high-quality, evidence-based care for those fortunate enough to access it. The setting up of these services has often been driven by clinicians who have also evaluated these service in relation to a number of outcomes, including compliance,[38,39] refracture rates[40] and costs of running the service.[41]

A randomised controlled trial undertaken by Ganda et al. in Sydney, Australia, provided important information on longer-term compliance and persistence with treatment for osteoporosis.[38] Whilst treatment initiation was under the umbrella of a well-established secondary fracture prevention service, the study explored the role of follow-up comparing review undertaken in a secondary fracture prevention programme versus follow-up by a general practitioner. Outcome measures were extracted from pharmaceutical claims data and demonstrated no significant differences in compliance between groups. This has implications for the operational aspects of secondary fracture prevention services including costs as it seems that after identification, investigation and initiation of treatment within a secondary fracture prevention programme, longer-term care can transition back to the general practitioner.

The optimal model of care is challenging to define and needs to take into account the evidence as well as what is practical and feasible from a local context including giving consideration to the challenges of delivering services to regional and rural centres in Australia.[42] Evidence supports enrolment of patients into 'programmes of care' as opposed to the traditional model of GP-based care but cost and sustainability remain ongoing concerns.[43–47]

In Western Australia, Inderjeeth and colleagues evaluated the impact of a hospital-based fracture liaison service both on refracture rate and cost of the service.[41,48] The results suggest that the rate of refracture compared to the retrospective cohort and other tertiary hospitals was reduced by between 9.2% and 10.2% equating to cost savings of approximately $750,168–$810,400/1000 patient years in the first year.

There are mounting data to support fracture liaison services or organised 'programmes of care' in relation to cost-effectiveness. However, in fragmented health-care systems, the investor in the service may not be the beneficiary of the potential gains as is the case when investing in osteoporosis services in the community to reduce and prevent hospitalisation for fracture in Australia.[49–52]

Moving to a State-Based Approach

New South Wales appears to be the first State with a cohesive plan to roll out fracture liaison services across all its local health districts. The approach has been many years in the planning and has been driven largely by the NSW Agency for Clinical Innovation[53,54] and more recently through an NSW Ministry of Health Initiative – Leading Better Value Care Program.[55] The Agency for Clinical Innovation's Musculoskeletal Network has systematically identified the issue at a State level, developed and evaluated potential solutions and released a model of care in 2011.[53] Implementation of the model is sufficiently flexible to accommodate the challenge of delivering fracture liaison services to rural and remote parts of New South Wales as well as inner city high population density areas.

One of the early adopter hospitals in NSW evaluated the impact of its fracture liaison service using a historical cohort study comparing rates to another hospital of similar size without a fracture liaison service.[47] The fracture liaison service is proactive in identifying people presenting to the hospital with a low-trauma fracture and contacts people by letter with an appointment to the clinic. Those who attend the clinic are assessed for future fracture risk and provided with dietary and lifestyle education as well as offered treatment for osteoporosis as and when indicated. Importantly, the study took an intention to treat approach and so included people who were contacted but did not attend the clinic. After taking into account differences in baseline characteristics between the two hospital populations, the study showed a 30% reduction in any refracture and a 40% reduction in major refractures (hip, spine, pelvis, femur or humerus) in favour of the site with the fracture liaison service.

The number of hospitals with a fracture liaison service in New South Wales is increasing. To facilitate the sharing of ideas and learnings, the ACI has published a Re-Fracture Prevention Service Directory which provides a list of all hospitals currently providing secondary fracture prevention services across Australia's most populous State.[54] The number of services in recent years has increased with more on the way. The majority are hospital based, although some have sought to integrate or work in partnership with primary care networks and general practitioners.

Following evaluation of the model of care, all state-funded health districts have been directed by the NSW Ministry of Health to implement the model of care in 2017/18 financial year. Key to the implementation is the availability of a fracture liaison coordinator. It is anticipated that activity-based funding will cover the costs of the roll out. There will undoubtedly be challenges in achieving this aim which to a degree overlooks the role of the general practitioner.

Only time will tell as to whether the anticipated reduction in fracture rate is observed. If successful, it is hoped that other States might take a lead from NSW and roll out similar services in other States.

HIP FRACTURE CARE – UNITED WE STAND

Hip fracture is a devastating injury for an older person and is associated with significant pain, loss of function and loss of independence. It also carries an excess mortality associated with the injury.[56,57] Frustratingly, almost 50% of people who fracture their hip will have sustained a previous low-trauma fracture, yet only around 20% will have been offered treatment for osteoporosis.[58–61] This is despite strong evidence to support secondary fracture prevention and the availability of doctors with an interest in the area, support for bone mineral density testing and the availability of a range of effective treatments in both countries.

ANZ Hip Fracture Registry

The Australian and New Zealand Hip Fracture Registry was formed in 2012. The Registry is based on the UK National Hip Fracture Database and to date has released five facility level audits and two patient level audits. The design and development of the Registry has been incremental. In September 2014, the National Health and Medical Research Council endorsed Australian and New Zealand Hip Fracture Guideline was released.[62] In Australia, the Australian Commission on Safety and Quality in Health Care has carriage of national standards. In 2016, the Hip Fracture Care Clinical Care Standard was released.[63] This document was developed in partnership with the New Zealand Health Quality and Safety Commission. Quality Statement number 6 specifically addresses the issue of secondary fracture prevention – 'Before a patient with a hip fracture leaves hospital, they are offered a falls and bone health assessment, and management plan based on this assessment, to reduce the risk of another fracture'. The quality statement is accompanied by measureable indicators and in the 2017 ANZ Hip Fracture Registry reports 31% and 16% of patients in

New Zealand and Australia, respectively, were receiving treatment for osteoporosis at discharge from the acute setting (Fig. 4.1).[64] These figures are at best disappointing. Some of these patients have low vitamin D levels and may require a period of replacement before offering definitive treatment for osteoporosis; however, even accounting for this, the figures suggest that care could and should be better. Hip fracture patients are a captive audience, sitting in hospital beds as they recover from a serious injury. However, in Australia there are challenges to initiating a treatment for a chronic disease in the hospital setting. The agreement between Commonwealth and State governments is such that hospital activity is funded through State health budgets and there is a reluctance to support the initiation in hospitals of treatments for a chronic disease when the treatments are perceived to be costly.

Fracture liaison services clearly have a role in the systematic identification of patients with a low-trauma fracture so as to ensure that these people are put on a pathway that leads to appropriate investigation and treatments. Whilst the identification of an at-risk population can be done in the acute setting, the follow-up can be community or out-patient based. The Australian and New Zealand Hip Fracture Registry specifically asks about the availability of fracture liaison services at hospitals operating on hip fracture patients. Over the past 5 years, there has been an increase in the number of hospitals reporting the availability of this service from 15% in 2013 to 33% in 2017 (Table 4.1). However, a number of sites report that this is predominantly targeting hip fracture patients and not all low-trauma fracture patients. The availability of publicly funded osteoporosis clinics has increased from 35% in 2013 to 40% in 2017, whilst little is known about how many private services are available and how many people access these services.

Quality statement 7 of the Hip Fracture Clinical Care Standard highlights the need to provide patients with a written plan that addresses future falls and fracture risk. This plan is separate to any discharge summary sent from the hospital to the local general practitioner. The intent is to empower patients to ensure that they play an active role in future fracture prevention. The latest report shows that provision of this information is low at 27% and highlights a missed opportunity in preventing the next fracture.

SUMMARY

Osteoporosis is a chronic disease which is amenable to intervention that can reduce fractures, pain and disability and preserve independence. A low-trauma fracture provides a signal to the person and the health system that there is an issue, and that steps are required to reduce falls risk and assess and manage bone health. After years of inertia and well-meaning but largely ineffective rhetoric, it appears that secondary fracture prevention is now on the radar of health departments and governments in Australia and New Zealand. New Zealand appears to have reached a tipping point with evidence of a national drive to roll out fracture

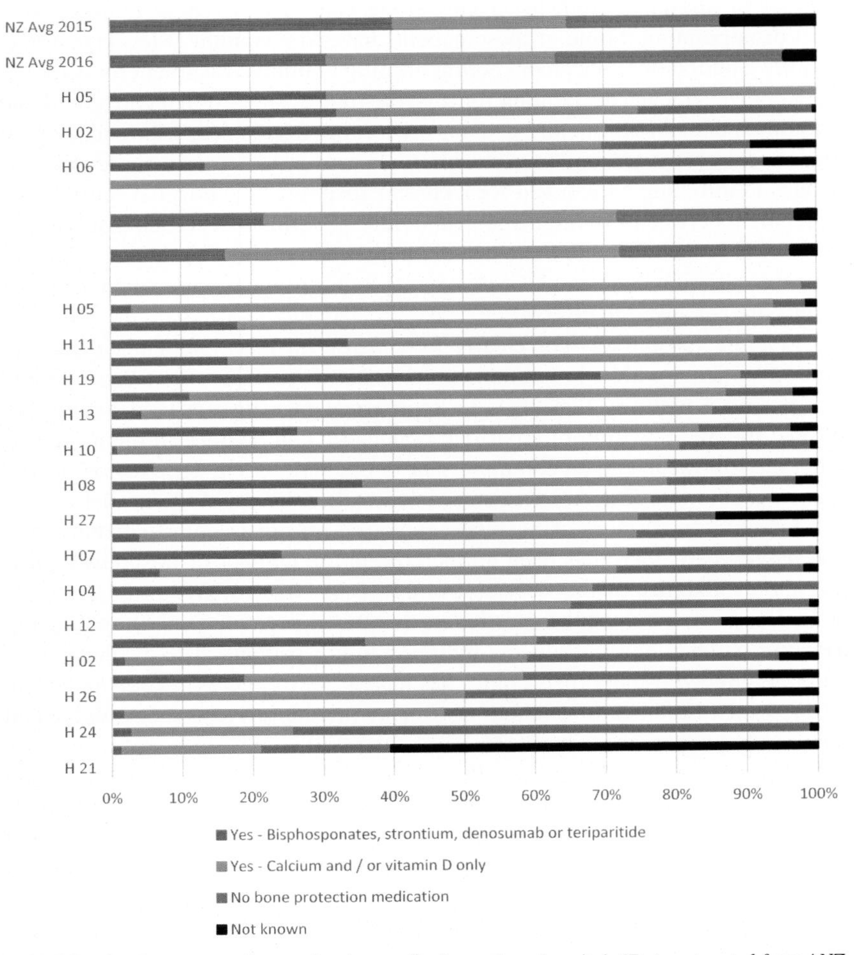

FIGURE 4.1 Bone protection medication on discharge from hospital. *(Data extracted from ANZ Hip Fracture Registry Report 2017.)*

liaison services across the country and over a relatively short period of time. Over the coming years, the success or otherwise of the approach will be evaluated, primarily by monitoring national fracture rates.

Momentum continues to gather in Australia towards a national approach to secondary fracture prevention. The size of the country and the fragmented approach to funding and delivery of health services creates a significant challenge. Whilst harder and slower to move an agenda forward, it is more likely that a national collaborative approach with a clear strategic intent will deliver sustainable change. State-led models are emerging that have a chance to implement

TABLE 4.1 Availability of Services for Secondary Fracture Prevention Over Time

	2013 (n = 116)	2014 (n = 117)	2015 (n = 120)	2016 (n = 121)	2017 (n = 120)
Fracture liaison service	15%	20%	21%	25%	33%
Access to a public falls clinic	41%	43%	57%	64%	58%
Access to a public osteoporosis clinic	35%	32%	40%	48%	40%
Access to a public falls and bone health clinic	16%	15%	18%	17%	16%
Access to a public orthopaedic clinic	72%	90%	91%	90%	89%
Routine provision written information on treatment and care after hip fracture	n/a	27%	41%	38%	39%
Routine provision of individualised written information on prevention of future falls and fractures	n/a	n/a	27%	27%	27%

Data extracted from ANZ Hip Fracture Registry Report 2017.

care in the future and ultimately deliver better outcomes for its population and reduced costs to health services. More work is also required from a public health perspective to increase awareness of both osteoporosis and falls and the media have a role to play in increasing awareness in the general population.

In both Australia and New Zealand, the next few years are critical to moving the secondary fracture prevention agenda forward. It is with hope and anticipation that the benefits of systematised care will be realised over a short period.

ACKNOWLEDGEMENTS

I would like to acknowledge Paul Mitchell in sharing documents with me that have assisted with the writing of this chapter.

REFERENCES

1. Australian Bureau of Statistics. *Population clock*. 2016. Available from: http://www.abs.gov.au/AUSSTATS/abs@.nsf/Web+Pages/Population+Clock?opendocument&ref=HPKI.
2. https://en.wikipedia.org/wiki/Australian_Bureau_of_Statistics Australian Bureau of Statistics.
3. Australian Institute of Health and Welfare. *Hospitalisation for osteoporosis*. 2016. Available from: http://www.aihw.gov.au/osteoporosis/hospitalisation-and-treatment/#table1.
4. Australian Institute of Health and Welfare. *Osteoporosis*. 2016. Available from: http://www.aihw.gov.au/osteoporosis/.
5. Watts JJ, Abimanyi-Ochom J, Sanders KM. *Osteoporosis costing all Australians A new burden of disease analysis – 2012 to 2022*. Glebe: Osteoporosis Australia; 2013.
6. Wikipedia. *New Zealand*. Available from: https://en.wikipedia.org/wiki/New_Zealand.
7. Wikipedia. *Politics of New Zealand*. Available from: https://en.wikipedia.org/wiki/Politics_of_New_Zealand.
8. Wikipedia. *Health Care in New Zealand*. Available from: https://en.wikipedia.org/wiki/Health_care_in_New_Zealand.
9. Brown P, McNeill R, Radwan E, Willingale J. *The burden of osteoporosis in New Zealand: 2007–2020*. Auckland: School of Population, Health University of Auckland; 2007.
10. Barrack CM, McGirr EE, Fuller JD, Foster NM, Ewald DP. Secondary prevention of osteoporosis post minimal trauma fracture in an Australian regional and rural population. *Aust J Rural Health* 2009;**17**(6):310–5.
11. Bliuc D, Ong CR, Eisman JA, Center JR. Barriers to effective management of osteoporosis in moderate and minimal trauma fractures: a prospective study. *Osteoporos Int* 2005;**16**(8):977–82.
12. Gunathilake R, Epstein E, McNeill S, Walsh B. Factors associated with receiving anti-osteoporosis treatment among older persons with minimal trauma hip fracture presenting to an acute orthogeriatric service. *Injury* 2016;**47**(10):2149–54.
13. Parker D. An audit of osteoporotic patients in an Australian general practice. *Aust Fam Physician* 2013;**42**(6):423–7.
14. Wong PK, Spencer DG, McElduff P, Manolios N, Larcos G, Howe GB. Secondary screening for osteoporosis in patients admitted with minimal-trauma fracture to a major teaching hospital. *Intern Med J* 2003;**33**(11):505–10.
15. Kimber CM, Grimmer-Somers KA. Evaluation of current practice: compliance with osteoporosis clinical guidelines in an outpatient fracture clinic. *Aust Health Rev* 2008;**32**(1):34–43.
16. Jenkins M, Smith P, Scarvell J. Osteoporosis: the forgotten diagnosis? A retrospective audit of patients presenting with fractures for osteoporosis management. *MSJA* 2011;**3**(2):12–7.

17. Kelly AM, Clooney M, Kerr D, Ebeling PR. When continuity of care breaks down: a systems failure in identification of osteoporosis risk in older patients treated for minimal trauma fractures. *Med J Aust* 2008;**188**(7):389–91.
18. Otmar R, Henry MJ, Kotowicz MA, Nicholson GC, Korn S, Pasco JA. Patterns of treatment in Australian men following fracture. *Osteoporos Int* 2011;**22**(1):249–54.
19. Inderjeeth CA, Glennon D, Petta A. Study of osteoporosis awareness, investigation and treatment of patients discharged from a tertiary public teaching hospital. *Intern Med J* 2006;**36**(9):547–51.
20. Myers TA, Briffa NK. Secondary and tertiary prevention in the management of low-trauma fracture. *Aust J Physiother* 2003;**49**(1):25–9.
21. Stracey-Clitherow H, Bossley C. Osteoporosis intervention by New Zealand orthopaedic units: a multi-centre audit. *J Bone Joint Surg Br* 2009;**91-B**(Suppl. II):342.
22. Stracey-Clitherow HD, Bossley CJ. *Osteoporosis intervention by New Zealand orthopaedic departments in patients with fragility fractures: a multi-centre audit.* Wellington. 2009.
23. Mitchell PJ, Cornish J, Milsom S, Hayman L, Seibel MJ, Close J, et al. BoneCare 2020: a systematic approach to hip fracture care and prevention for New Zealand. In: *3rd fragility fracture network congress 2014; 4–6 September 2014; Madrid, Spain.* 2014.
24. Eisman J, Clapham S, Kehoe L, Australian BoneCare Study. Osteoporosis prevalence and levels of treatment in primary care: the Australian BoneCare Study. *J Bone Miner Res* 2004;**19**(12):1969–75.
25. Chen JS, Hogan C, Lyubomirsky G, Sambrook PN. Management of osteoporosis in primary care in Australia. *Osteoporos Int* 2009;**20**(3):491–6.
26. Lai MMY, Ang WM, McGuiness M, Larke AB. Undertreatment of osteoporosis in regional Western Australia. *Australas J Ageing* 2012;**31**(2):110–4.
27. Duque G, Lord SR, Mak J, Ganda K, Close JJ, Ebeling P, et al. Treatment of osteoporosis in australian residential aged care facilities: update on consensus recommendations for fracture prevention. *J Am Med Dir Assoc* 2016;**17**(9):852–9.
28. Wong YYE, Flicker L, Draper G, Lai MMY, Waldron N. Hip fractures among Indigenous Western Australians from 1999 to 2009. *Intern Med J* 2013;**43**(12):1287–92.
29. Mitchell PJ, Ganda K, Seibel MJ. *Australian and New Zealand bone and mineral society position paper on secondary fracture prevention programs.* Available from: https://www.anzbms.org.au/downloads/ANZBMSPositionPaperonSecondaryFracturePreventionApril2015.pdf.
30. Augat P, Weyand D, Panzer S, Klier T. Osteoporosis prevalence and fracture characteristics in elderly female patients with fractures. *Arch Orthop Trauma Surg* 2010;**130**(11):1405–10.
31. Baba T, Hagino H, Nonomiya H, Ikuta T, Shoda E, Mogami A, et al. Inadequate management for secondary fracture prevention in patients with distal radius fracture by trauma surgeons. *Osteoporos Int* 2015;**26**(7):1959–63.
32. Australian Government: Australian Institute of Health and Welfare. *Health priority areas Canberra.* 2002. Available from: http://www.aihw.gov.au/national-health-priority-areas/.
33. National Health Priority Action Council (NHPAC). In: Australian Government Department of Health and Ageing, editor. *National service improvement framework for osteoarthritis, rheumatoid arthritis and osteoporosis.* 2006. Canberra.
34. National Arthritis and Musculoskeletal Conditions Advisory Group. In: Australian Government Department of Health and Ageing, editor. *A National action plan for osteoarthritis, rheumatoid arthritis and osteoporosis.* 2005. Canberra.
35. Osteoporosis National Action Plan Working Group. *Osteoporosis national action plan 2016.* Sydney: Australian Institute for Musculoskeletal Science, Australian and New Zealand Bone and Mineral Society, Garvan Institute of Medical Research, Monash University, Move, Osteoporosis Australia, University of Sydney, University of Melbourne, UNSW Australia; 2016.

36. Primary Health Care Advisory Group Final Report. In: Australian Government Department of Health, editor. *Better outcomes for people with chronic and complex health conditions*. 2015. Canberra.
37. Bahl S, Coates PS, Greenspan SL. The management of osteoporosis following hip fracture: have we improved our care? *Osteoporos Int* 2003;**14**(11):884–8.
38. Ganda K, Schaffer A, Pearson S, Seibel MJ. Compliance and persistence to oral bisphosphonate therapy following initiation within a secondary fracture prevention program: a randomised controlled trial of specialist vs. non-specialist management. *Osteoporos Int* 2014;**25**(4):1345–55.
39. Inderjeeth CA, Glennon DA, Poland KE, Ingram KV, Prince RL, Van VR, et al. A multimodal intervention to improve fragility fracture management in patients presenting to emergency departments. *Med J Aust* 2010;**193**(3):149–53.
40. Ganda K, Schaffer A, Seibel MJ. Predictors of re-fracture amongst patients managed within a secondary fracture prevention program: a 7-year prospective study. *Osteoporos Int* 2015;**26**(2):543–51.
41. Inderjeeth C, Raymond W, Geelhoed E, Briggs A, Briffa K, Oldham D, et al. Hospital based Fracture Liaison Service reduces re-fracture rate and is cost-effective and cost saving. *J Bone Miner Res* 2016;**31**(Suppl. 1).
42. Fraser S, Wong PK. Secondary fracture prevention needs to happen in the country too: the first two and a half years of the Coffs Fracture Prevention Clinic. *Aust J Rural Health* 2017;**25**(1):28–33.
43. Dell R. Fracture prevention in Kaiser Permanente Southern California. *Osteoporos Int* 2011;**22**(Suppl. 3):457–60.
44. Hawley S, Javaid MK, Prieto-Alhambra D, Lippett J, Sheard S, Arden NK, et al. Clinical effectiveness of orthogeriatric and fracture liaison service models of care for hip fracture patients: population-based longitudinal study. *Age Ageing* 2016;**45**(2):236–42.
45. Huntjens KM, van Geel TA, van den Bergh JP, van Helden S, Willems P, Winkens B, et al. Fracture liaison service: impact on subsequent nonvertebral fracture incidence and mortality. *J Bone Joint Surg Am* 2014;**96**(4):e29.
46. Lih A, Nandapalan H, Kim M, Yap C, Lee P, Ganda K, et al. Targeted intervention reduces refracture rates in patients with incident non-vertebral osteoporotic fractures: a 4-year prospective controlled study. *Osteoporos Int* 2011;**22**(3):849–58.
47. Nakayama A, Major G, Holliday E, Attia J, Bogduk N. Evidence of effectiveness of a fracture liaison service to reduce the re-fracture rate. *Osteoporos Int* 2016;**27**(3):873–9.
48. Inderjeeth CA, Raymond W, Geelhoed E, Briggs A, Briffa K, Oldham D, et al. THU0465 fracture liaison service reduces re-fracture rate, cost-effective and cost saving in Western Australia. *Ann Rheum Dis* 2016;**75**(Suppl. 2):360–1.
49. Cooper MS, Palmer AJ, Seibel MJ. Cost-effectiveness of the Concord Minimal Trauma Fracture Liaison service, a prospective, controlled fracture prevention study. *Osteoporos Int* 2012;**23**(1):97–107.
50. McLellan AR, Wolowacz SE, Zimovetz EA, Beard SM, Lock S, McCrink L, et al. Fracture liaison services for the evaluation and management of patients with osteoporotic fracture: a cost-effectiveness evaluation based on data collected over 8 years of service provision. *Osteoporos Int* 2011;**22**(7):2083–98.
51. Sander B, Elliot-Gibson V, Beaton DE, Bogoch ER, Maetzel A. A coordinator program in post-fracture osteoporosis management improves outcomes and saves costs. *J Bone Joint Surg Am* 2008;**90**(6):1197–205.
52. Solomon DH, Patrick AR, Schousboe J, Losina E. The potential economic benefits of improved post-fracture care: a cost-effectiveness analysis of a fracture liaison service in the us health care system. *J Bone Miner Res* 2014;**29**(7):1667–74.

53. New South Wales Agency for Clinical Innovation Musculoskeletal Network. *NSW model of care for osteoporotic refracture prevention.* 2011.
54. New South Wales Agency for Clinical Innovation Musculoskeletal Network. *NSW osteoporosis re-fracture prevention service directory: version 9 - September 2016.* Chatswood, NSW: New South Wales Agency for Clinical Innovation; 2016.
55. NSW Ministry of Health Leading Better Value Care Program. http://eih.health.nsw.gov.au/bvh/about/leading-better-value-care-program.
56. Hindmarsh D, Loh M, Finch CF, Hayen A, Close JCT. Effect of comorbidity on relative survival following hospitalisation for fall-related hip fracture in older people. *Australas J Ageing* 2014;**33**(3):E1–7.
57. Hindmarsh D, Hayen A, Finch C, Close J. Relative survival after hospitalisation for hip fracture in older people in New South Wales, Australia. *Osteoporos Int* 2009;**20**(2):221–9.
58. Edwards BJ, Bunta AD, Simonelli C, Bolander M, Fitzpatrick LA. Prior fractures are common in patients with subsequent hip fractures. *Clin Orthop Relat Res* 2007;**461**:226–30.
59. Gallagher JC, Melton LJ, Riggs BL, Bergstrath E. Epidemiology of fractures of the proximal femur in Rochester, Minnesota. *Clin Orthop Relat Res* 1980;**150**:163–71.
60. McLellan A, Reid D, Forbes K, Reid R, Campbell C, Gregori A, et al. *Effectiveness of strategies for the secondary prevention of osteoporotic fractures in Scotland (CEPS 99/03).* NHS Quality Improvement Scotland; 2004.
61. Port L, Center J, Briffa NK, Nguyen T, Cumming R, Eisman J. Osteoporotic fracture: missed opportunity for intervention. *Osteoporos Int* 2003;**14**(9):780–4.
62. Australian and New Zealand Hip Fracture Registry (ANZHFR) Steering Group. *Australian and New Zealand guideline for hip fracture care: improving outcomes in hip fracture management of adults.* Sydney: Australian and New Zealand Hip Fracture Registry Steering Group; 2014.
63. Australian Commission on Safety and Quality in Health Care, Health Quality & Safety Commission New Zealand. *Hip fracture care clinical care standard.* Sydney. 2016.
64. Andrade SE, Majumdar SR, Chan KA, Buist DS, Go AS, Goodman M, et al. Low frequency of treatment of osteoporosis among postmenopausal women following a fracture. *Arch Intern Med* 2003;**163**(17):2052–7.

Chapter 5

Fracture Liaison Services – Canada

Victoria Elliot-Gibson[1,2,3], Joanna Sale[1,4,5], Ravi Jain[3], Earl Bogoch[2,5,6]
[1]Musculoskeletal Health and Outcomes Research, Li Ka Shing Knowledge Institute, St. Michael's Hospital, Toronto, ON, Canada; [2]Mobility Program. St. Michael's Hospital, Toronto, ON, Canada; [3]Ontario Osteoporosis Strategy, Osteoporosis Canada, Toronto, ON, Canada; [4]Institute of Health Policy, Management & Evaluation, University of Toronto, Toronto, ON, Canada; [5]Li Ka Shing Knowledge Institute, St. Michael's Hospital, Toronto, ON, Canada; [6]Department of Surgery, University of Toronto, Toronto, ON, Canada

INTRODUCTION

With 10 provinces, 3 territories and almost 10 million square kilometres, Canada is the second largest country on earth but has only 36 million residents, with a low population density of approximately 3.9 people per square kilometre.[1,2] The geographic distribution of the population is a challenge for the delivery of health care, with over 60% of the population in two provinces, Ontario and Quebec, and 66% of the population within 100 kilometres of the Canada–United States border.[2] In 2017, 30% of the Canadian population were aged ≥50 years with a life expectancy of 79 years for men and 83 years for women.[3,4] Age structure forecasts project rapid growth of the Canadian senior population over the next two decades. It is estimated the proportion of the population aged ≥65 will increase to approximately 23% compared to 16% in 2017.[3,5] Age is a strong predictor of fracture[6] and the aging of the Canadian population has important implications for society and the economy.

In 2011, an estimated 131,443 fractures in Canada could be attributed to impaired bone strength, resulting in 64,884 acute care admissions and 983,074 acute hospital days.[7] For men and women aged 50 to 99, 29,902 sustained a hip fracture related to osteoporosis which represented 22.7% of fractures. The average cost per hip fracture patient was $63,649, much greater than the cost per wrist fracture ($8681), vertebra ($26,960) and shoulder ($15,862).[7] Overall, the total economic burden of fragility fractures in Canada for fiscal year 2010/2011 was approximately $4.6 billion Canadian dollars in year 2014 values ($4.1 billion US dollars in year 2014 values; $3.1 billion Euros in year 2014 values).

Secondary Fracture Prevention. https://doi.org/10.1016/B978-0-12-813136-7.00005-3

Beyond the economic burden and the increased risk of future fracture[8–10] and mortality,[11–13] fragility fractures negatively impact Canadians' health related quality of life,[13–15] independence and function.[13,16] The challenge of overcoming the post fragility fracture care gap has been extensively studied and published through randomized and nonrandomized trials in Canada[17] and is the focus of intensive efforts by the Canadian community promoting Fracture Liaison Services (FLS).[18–24]

CANADIAN HEALTH CARE SYSTEM

Canada has a publicly funded health-care system principally delivered by the 10 provinces and 3 territories,[25] which have constitutional responsibility for this service. The Federal Government of Canada, through the Canada Health Transfer, provides funds and tax transfers to the provinces and territories to support hospitals, physicians and community-based services. Each province and territory independently determines what services are medically necessary and covered by their provincial health-care system. Physicians' services are funded mainly through a fee-for-service schedule. Some provinces cover costs associated with long-term care, home care, rehabilitation and prescription medications for those aged ≥65 years and also for those aged <65 years who live in long-term care homes, receive disability or social assistance income support or have high prescription drug costs relative to their income. Otherwise, Canadians pay out of pocket or through private insurance offered by employers or purchased privately for prescription medications, physical therapy, dental care, vision care, mobility devices and other health-care services.

ACCESS TO BONE MINERAL DENSITY TESTING IN CANADA

The current Osteoporosis Canada (OC) clinical practice guidelines for the diagnosis and management of osteoporosis have recommendations for bone mineral density (BMD) testing for women and men ≥ 65 years, postmenopausal women and men aged 50 to 64 and those under age 50, with internationally recognized risk factors for fracture.[26] However, each province and territory have eligibility criteria for BMD testing.

In Ontario, the provincial insurance plan funds an annual BMD test for patients at high risk of osteoporosis, including those with prevalent fragility fracture, premature menopause, high dose steroid therapy and other prevalent risk factors.[27] In Nova Scotia, the indications for reimbursed BMD testing follow OC's 2010 Guidelines for the initial BMD test.[28] In Alberta, BMD is funded for patients aged ≥ 50 years and may be performed once in every 2 years.[29] For more frequent testing, the physician must enter specific indications on requisition. For patients <50 years old, referral by a specialist is required.[29]

ACCESS TO PHARMACOTHERAPY FOR FRACTURE PREVENTION IN CANADA

Prescription medications available in Canada to lower osteoporotic fracture risk include alendronate, risedronate, zoledronic acid, denosumab, raloxifene, estrogen and teriparatide. Etidronate remains available but is rarely prescribed. Generic preparations are available for all the bisphosphonates. Each province and territory has eligibility criteria for pharmacotherapy funding through provincial/territorial public drug plans and continuing processes for updating the criteria. Availability of medications in each province and territory is updated on OC's website.[30]

FRACTURE RISK ASSESSMENT TOOLS IN CANADA

The Canadian Association of Radiologists/Osteoporosis Canada, CAROC, is a Canadian fracture risk assessment tool.[26] CAROC 2010 stratifies women and men, aged ≥50 years in three bands of risk for future fragility fracture over 10 years: low (<10%), moderate (10%–20%) and high (>20%). It incorporates several risk factors including age, sex and femoral neck T-score. Prior fragility fractures and glucocorticoid use each raise the derived risk level to the next category.[26] Guidelines recommend that pharmacologic therapy be offered to patients at high absolute risk of future fracture. In addition, patients at moderate risk of future fracture should be considered for pharmacotherapy based on a bone health assessment to identify additional risk factors that are not considered in the risk assessment system, according to the clinician's judgement. The CAROC tool was validated in the Canadian population[31] and is available online and for mobile use at: http://www.osteoporosis.ca/multimedia/FractureRiskTool/index.html#/Home.

A Canadian version of the widely used FRAX tool constructed using Canadian national hip fracture data is applicable for use, also including a mobile application, with or without the femoral neck BMD T-score.[32] https://www.sheffield.ac.uk/FRAX/tool.aspx?country=19. In Canada, intervention thresholds using a fixed FRAX probability for major osteoporosis fracture is >20%, but no set threshold percentage for hip fracture has been determined, whereas other organisations use a 3% hip fracture risk intervention threshold.[33]

THE ROLE OF OSTEOPOROSIS CANADA IN SUPPORTING THE IMPLEMENTATION OF QUALITY FLS IN CANADA[34]

OC is the national nonprofit organisations serving people who have, or are at risk of, osteoporosis. Since February 2010, OC has been focused on 'highest risk' patients, defined as those who have sustained a fragility fracture.

Released in March 2011, OC's White paper, *Towards a fracture-free future*,[35] was a call to action for provincial governments to implement coordinated post-fracture care programmes with case managers to effectively identify and manage Canadians who present with fragility fractures.

The systematic review by Sale et al., in 2011[17] and Ganda et al., in 2013,[36] further shaped OC's definition of an effective post-fracture care programme. These two landmark studies arrived at the same salient conclusions: successful post-fracture care models require dedicated personnel and a critical intensity. Effective models are those interventions which conduct investigations and/or initiation of osteoporosis treatment within the programme.

In March 2013, the national Board of Directors directed the organisations to focus its advocacy efforts on FLS according to criteria for the most effective interventions established in the Sale and Ganda articles.

OC has developed numerous tools and resources to support the implementation of effective FLS in Canada. These are found on the OC *FLS Hub* (http://www.osteoporosis.ca/fls/), and include:

- The OC official definition of FLS (Table 5.1).
- An *FLS Toolkit*, with useful tools and templates to assist new or aspiring FLS teams in the design and implementation of an FLS model.
- *Essential Elements of FLS*, a document outlining the minimal essential components of a successful FLS.
- The OC *FLS Registry* showcasing Canadian FLSs meeting all eight of the Essential Elements. As of August, 2018, 47 FLS programmes are listed on the *FLS Registry*, many fewer than are needed in Canada.
- *Key indicators for Canadian FLSs* to support them in continuous quality improvement (CQI).
- *FLS Quality Standards*, a set of high-level standards to which all FLS should aspire. This document has been endorsed by eight national professional organisations.
- *FLS Works!* webinar series.
- A free consultation service for new/aspiring Canadian FLS.
- The *FLS Network*, Canada's largest network of health-care professionals and health-care administrators interested in quality of FLSs. Members are kept informed of new developments with *Liaison* newsletters.
- OC is conducting Canada's first ever national FLS audit in 2018.

TABLE 5.1 Osteoporosis Canada's Definition of Fracture Liaison Service

Identification	• Systematically and proactively identifies patients aged 50 years and older presenting to a hospital with a new fragility fracture and/or with a newly reported vertebral fracture;
Investigation	• Organizes appropriate investigations to determine the patient's fracture risk;
Initiation	• Facilitates the initiation of appropriate osteoporosis medications.

OC provides ongoing support for FLS champions in the provinces, most of whom have not yet succeeded in establishing a local FLS. OC coordinates and helps draft business cases, presentations and other documents for meetings with key decision makers. National FLS meetings, the FLS Summit in 2014 and the FLS Forum in 2017, were held to bring together key stakeholders from each province to help advance the FLS cause in their jurisdictions.

ONTARIO PROVINCIAL FRACTURE LIAISON SERVICE

In the province of Ontario, with a population of 14 million,[1] health care is managed through 14 Local Health Integration Networks (LHINs) funded by the Ontario Ministry of Health and Long-Term Care. The LHINs are responsible for hospitals, long-term care homes, Community Care Access Centres, Community Support Services, Community Health Centres and Addictions and Mental Health Agencies. Physician remuneration is by the provincial government, through the Ontario Health Insurance Plan, largely on a fee-for-service basis.[37]

In 2003, a report entitled the 'The Osteoporosis Action Plan: An Osteoporosis Strategy'[38] outlined a framework to provide access to best practices in osteoporosis prevention and management in Ontario. This plan focused on: health promotion, early detection and diagnosis of osteoporosis, better access to effective, evidence-based treatments, integrated post-fracture care, rehabilitation and osteoporosis management, self-management and falls prevention, evidence-based professional practice, research to develop new knowledge in osteoporosis prevention, diagnosis and management, and leadership and monitoring to ensure effective implementation of the plan. This strategy was adopted by the Ontario government in February 2005 with an annual commitment of 5 million (Canadian dollars in year 2005 values, $4 million US dollars in year 2005 values, $3 million Euros in year 2005 values). The Ontario Osteoporosis Strategy (OOS) is a patient-centred, multidisciplinary approach integrated across health-care sectors, purposed to reduce morbidity, mortality and costs resulting from osteoporotic fractures. The Strategy has multiple stakeholders including OC and is structured to achieve its objectives by changing practices at the health systems level (*Fracture Prevention* priority), educating health-care practitioners (*Professional Education and Outreach* priority) and educating and empowering patients (*Patient Education and Self-Management* priority).[39]

Ontario Wide Fracture Screening and Prevention Programme

The Ontario government funded Fracture Screening and Prevention Programme (FSPP) was initiated under the OOS and is implemented and managed by OC. FSPP was established on a scalable model focused on quality improvement within the outpatient orthopaedic environment anticipating expansion to cover all hospital sites in Ontario. The FSPP is now operating in 36 medium and high-volume Ontario hospital fracture clinics and has evolved from a basic screening and education initiative, into a more intensive FLS which leverages clinical and diagnostic supports available at each site through Fracture Prevention Coordinators who are centrally trained and supervised. The role of the coordinator is to identify, assess, refer and educate fragility fracture patients regarding bone health and to be the liaison arranging individualized diagnostic and clinical follow-up appropriate for each patient according to OC guidelines. Patients consent to participate and approval from each hospitals research ethic board is obtained to permit the utilization of the data collected for quality improvement and research purposes.

Patient Identification

Currently, the coordinator screens men and women aged ≥50 years who present in outpatient fracture clinics with a fragility fracture of the wrist, elbow, shoulder, clavicle, vertebra, pelvis, proximal and distal femur, and tibia/fibula. Patients are excluded if they sustained a fracture greater than 1 year prior to identification, or are unable to or decline to participate. The FSPP has screened over 1.5 million patient visits since 2007 and enrolls 8000–8500 fracture patients annually across 36 sites. There are over 76,000 unique patients documented in the programme database. The programme aims to broaden the screening programme to cover more sites as more funding becomes available.

Education

The coordinator provides fracture patients with educational materials on fracture risk assessment, exercise and nutrition provided by OC. In addition, FSPP employs nine Regional Integration Leads who support and oversee the FSPP and develop regional partnerships with primary and community care providers to integrate post-fracture care pathways. They also develop bone health educational collaborations and help disseminate tools and resources designed for health-care professionals, patients and caregivers.

Iterative Processes of Programme Quality Improvement

From 2007 to 2011, the coordinator functioned in a relatively limited educational and liaison role, through targeted discussions with the patient and communications with their primary care provider (PCP). Based on early data demonstrating that the model employed by the FSPP produced only modest outcome improvements[40–42] a more intensive iteration of the FSPP was initiated at selected sites in 2011. This improved pathway is internally identified as the ‘BMD fast track’ sites, and the coordinator has the authority to arrange BMD testing on behalf of the orthopaedic surgeon or through a medical directive.

A comparison of patients managed in the original FSPP design or in the BMD ‘fast track’ innovation revealed that BMD ‘fast track’ design led to a significant improvement in treatment initiation for patients who were treatment naïve when screened (32% vs. 16% across all risk categories).[43]

Further improvements were noted once osteoporosis specialist referral pathways were established with a prescription rate at 47% (for 2016–17) for combined high and moderate risk patients and 70% for patients documented to be at high risk of future fracture.[44]

In addition, no evidence of gender inequity persisted after 'fast track' initiation in pharmacotherapy initiation, after adjusting for subsequent fracture risk, at 6-month follow-up 68.4% of women and 66.2% of men ($P>.05$) at high risk were treated within 6 months of screening.[45] This unique achievement, with both high risk males and females having similar rates of treatment initiation, is in contrast to the widely documented gender inequity reported from various jurisdictions.[46,47] We speculate that the systematic processes of FSPP neutralize the effect of endemic gender-related assumptions in delivery of health care in this domain.

Bone Mineral Density Testing

BMD test results and the calculated fracture risk assessment are reviewed by the coordinator who provides this information and the relevant guideline to the patient's PCP to promote a treatment decision.[43] After implementation of the BMD fast track programme improvement, patients were more likely to have a BMD test completed than previously (96% vs. 66%) and were more likely to visit their PCP to discuss the BMD results. In December 2017, the BMD fast track version of the FSPP was operational in 35 of 36 clinic sites.

Referrals for Bone Health Assessment and Pharmacotherapy

Patients at high risk of future fracture as well as patients who were receiving pharmacotherapy at the time of fracture can be referred to an osteoporosis specialist. The coordinator communicates with the patient's PCP and, where applicable, refers the patient to the osteoporosis specialist. The coordinator also contacts the patient or caregiver after a few weeks to determine whether recommended diagnostic and follow-up care was received. A key goal of the FSPP is to engage patients in self-management of bone health. To this end, the programme maintains a focus on continual improvement in patient uptake of attendance at consultations, recommended treatment and improvement of patient compliance.

Data Collection and Programme Evaluation

The coordinator screens and arranges referral through a live, interactive cloud-based database accessed using a secure network through a handheld device (tablet). Use of the electronic screening and referral database allows FSPP to integrate data quality checks within the screening process itself. It provides data for analysis in a CQI process.[48] For instance, validation checks are built-in so data entry errors or incorrect information, such as data outside a reference range, can be corrected in real time. The database also contains an algorithm which generates an automatic referral letter and routes patients to the appropriate care pathway based on patient response to the screening questions.

An example of quality improvement, enabled by the database, was accomplished regarding BMD reporting. Approximately 20% of BMD reports stated inappropriate low future fracture risk and did not increase the patients future fracture risk in light of the prevalent fragility fracture. In cases where discordance between reported and actual fracture risks (using CAROC/FRAX) was found, feedback was provided to the reporting physicians at BMD sites to correct the individual fracture risk report and to guide future reporting. This resulted in a 50% reduction in BMD reporting errors.[49]

Programme Strengths and Limitations

There are processes in place within the FSPP to identify, assess and correct care gaps along the intervention pathway. The programme is managed at the provincial level, where inter-hospital and regional comparisons are analysed for constructive feedback to improve programme performance. Resource allocation is periodically readjusted at a regional or provincial level to enhance efficiency. The FSPP is currently exploring opportunities to add telephone and telemedicine screening which can be cost-effective in remote and less-populated regions.

Patient data security is carefully maintained. Protocols which restrict data access and encrypt data capture as well as protect data transfer and storage are in place. Privacy and threat risk assessments are regularly conducted to ensure compliance with current standards. Only de-identified data are used for analysis and evaluation purposes and access to patient identifiers is restricted to the coordinator. FSPP obtains specific patient consent to enable linkage to the provincial patient databases for long-term tracking of patient status, including refracture rates.

The FSPP has limitations of which the most important is that the programme screens outpatient clinics and does not have resources to screen inpatients admitted at the FSPP sites. FPCs only screen those inpatients who return to fracture clinic for follow-up. An important priority for the FSPP is to marshal resources to screen inpatient fracture patients as soon as possible. The programme is developing processes to obtain patient lists and pursue screening by telephone to improve intervention rates for missed inpatients. The programme has exclusion criteria that may be excessively broad and also cannot screen patients without their consent. The coordinator is not an employee of the hospital(s) in which they work, and OC has contracts with each hospital allowing the coordinator to work in the hospital's clinical space. This requires special efforts to ensure that the coordinator is integrated into the fracture clinic team. The coordinator must also be processed through hospital security protocols and sign a confidentiality agreement before communicating with patients.

NOVA SCOTIA PROVINCIAL FRACTURE LIAISON SERVICES

A post-fracture care gap was documented in the province of Nova Scotia. In 2010, only 23% of hip fracture patients received pharmacotherapy to lower their risk of future fracture within 6 months of their fracture.[50] An FLS prototype, consistent with OC's 2010 clinical practice guidelines, was designed in 2010 to respond to these concerns.[26] Details of this programme are available online at: http://www.nshealth.ca/service-details/Fracture%20Liaison%20Service%20(FLS).

The Nova Scotia FLS prototype used the FLS at St. Michael's Hospital in Toronto[18,19] as the model, adapted to the Nova Scotia health-care system. The FLS nurse coordinator received training at St. Michael's Hospital.

The principal goal of the FLS is to increase the proportion of patients at high risk receiving first-line pharmacotherapy for fracture prevention. For optimal utilization of health-care resources, the FLS captures men and women, aged ≥50 years old, with a fragility fracture of the hip, vertebra, wrist, shoulder and pelvis. Systematic and proactive case finding is done in the hospital's orthopaedic inpatient wards and outpatient clinics. The FLS nurse has an expanded scope of practice and works independently with predetermined protocols which includes ordering BMD tests, or obtaining recent BMD test results, collecting the patient's clinical risk factors, and determine the patient's 10-year fracture risk using FRAX. The FLS nurse provides a detailed management plan to the patient's PCP, including a recommendation to initiate pharmacotherapy for patients deemed to be at high risk of fracture. The FLS recommends referral to a specialist only if indicated according to the 2010 OC guidelines.[26]

In the first year of the FLS prototype, implemented at Dartmouth General Hospital in February 2013, 66% of the identified patients were found to be at high risk.[51] At the end of the first year, 79% of hip fracture patients and 81% of the total high-risk patients had received a first-line pharmacotherapy within 6 months of their fracture (excluding patients lost to follow-up).[51]

Based on the success of this first FLS, a Nova Scotia provincial FLS programme was established. As of January 2018, three of the five orthopaedic hospitals in the province offer FLSs, both for inpatient and outpatient fracture patients.

The FLS programmes in Nova Scotia have been recognized as meeting OC's *Essential Elements of FLS*, are featured on OC's FLS Registry and have started to monitor their effectiveness utilizing the newly developed OC FLS indicators.[52] It is anticipated that FLS will be expanded to the remaining two orthopaedic hospitals of the province.

ALBERTA PROVINCIAL FRACTURE LIAISON SERVICES[53]

A post-fracture care gap was also documented in Alberta, where unpublished data suggest that ≤20% of hip fracture patients received assessment or adequate treatment for osteoporosis. Alberta Health Services' Bone and Joint Health Strategic Clinical Network is committed to implementing and managing Alberta's FLS programme. The first FLS opened at Edmonton Misericordia Community Hospital in June 2015, and there are five FLS sites in the province, with a sixth opening in January 2018, covering three zones, Edmonton, Central and Calgary Zones. Expansion to South Zone is on the horizon for 2018 with province-wide implementation at all surgical sites expected over the next few years.

Following the FLS medical algorithm, case management is undertaken by a registered nurse and a physician or team of physicians who will assess patients for appropriate pharmacotherapy. Pharmacists and specialists provide input for recommendations and management of complex cases.

All FLS sites in Alberta provide identification of men and women, ≥ 50 years of age who are present with a hip fracture in the inpatient setting. As per 2010 OC Guidelines,[26] those at risk undergo appropriate investigations and where appropriate, pharmacotherapy is initiated. Patients also receive a falls assessment and a comprehensive geriatric assessment. The FLS nurse will involve complementary health-care practitioners in areas such as dietetics, geriatrics, physiotherapy, pharmacy and others, as required.

The FLS nurse monitors patients for 1 year to assess barriers to compliance with the medication prescribed in addition to updating the PCP regarding test results. Prior to discharge from the FLS programme at 1 year, a final handover letter outlining the plan of care is provided.

Alberta's FLS programme maintains a provincial database. Quarterly reports are produced and FLS teams meet semiannually for education and networking for improved patient care.

INDEPENDENT FRACTURE LIAISON SERVICES IN CANADA

Although Ontario has an extensive provincially supported FLS network, there are several independent FLS programmes in the province providing identification, assessment and appropriate intervention to fracture patients.[54–57] The programme at St. Michael's Hospital is highlighted. British Columbia[58–60] and Quebec[61,62] provincial governments have fracture prevention guidelines and independent FLS programmes (Table 5.2). A full list of FLS programme across Canada is available at www.osteoporois.ca/fls.

Fracture Liaison Services at St. Michael's Hospital, Toronto, Ontario, Canada[18,19]

The FLS at St. Michael's Hospital, which is independent of the Provincial Ontario FSPP, is Canada's longest running FLS.

St. Michael's Hospital is a large urban university trauma hospital. This multidisciplinary, coordinator-based FLS was initiated by the Division of Orthopaedic Surgery in 2002 as a quality improvement programme to facilitate the identification, education, bone health assessment and appropriate treatment for inpatients and outpatients who have sustained a fragility fracture, and results of this programme have been previously reported.[18,19] Fig. 5.1 depicts the role of the coordinator. This FLS undergoes regular iterative programme modifications

TABLE 5.2 Independent FLS Programmes in Canada

Programme/ Province	Setting	Fracture Location(s)	Gender and Age	Programme Components	Outcomes
Fracture? Think Osteoporosis Program (FTOP), Hamilton, Ontario[54,55]	Inpatient rehabilitation unit at Hamilton Health Sciences Centre	Hip	Male and Females Age ≥ 50 years	Clinical nurse specialist educates patients on osteoporosis, falls, and appropriate nutritional and supplement intake. Refers patients back to PCP for BMD and pharmacotherapy. Informs PCP. Vitamin D and calcium initiated.	2008 cohort n = 110 Average age = 80.7 72% female BMD scheduled or performed: n = 54 (46%) On pharmacotherapy prior to fracture: n = 44 (40%) On pharmacotherapy 6 months follow-up: n = 82 (75%) 2011 cohort n = 97 Average age 82.3 73% female BMD scheduled or performed: n = 15 (14%) Pharmacotherapy prior to fracture: n = 28 (29%) On pharmacotherapy 6 months follow-up: n = 54 (56%)
University Health Network: Toronto Western Hospital[56]	Outpatient Fracture Clinic	All fracture locations (excluding hands, feet, skull and ankle)	Females ≥ 40 years Males ≥ 50 years	Coordinator identifies and educates patients on bone health, calcium and vitamin D, physical activity, falls prevention and osteoporosis and completes BMD testing. Using CAROC 2010, patients at high risk of future fracture are referred to a bone health specialist. Patients are followed up during their fracture clinic visit.	N/A

TABLE 5.2 Independent FLS Programmes in Canada—cont'd

Programme/ Province	Setting	Fracture Location(s)	Gender and Age	Programme Components	Outcomes
Geriatric Hip Fracture Program, Halton Healthcare Services (HHS), Oakville, Ontario[57]	Inpatients	Hip	Male and Females Age≥50 years	Geriatrician see's all hip fracture patients during their inpatient stay for appropriate evaluation and treatment of their osteoporosis prior to discharge	N/A
Peace Arch Hospital in the Fraser Health Region[59], British Columbia	Outpatient orthopaedic clinic	Wrist, shoulder, pelvis, hip and vertebra	Male and Females Age≥50 years	Nurse practitioner (NP) identifies, investigates and treats patients based on future fracture risk scoring using FRAX	N=122 65% completed a BMD test 81% of high risk patients started on pharmacotherapy/referred to osteoporosis consultant 100% of patients already on pharmacotherapy assessed for treatment change
St. Paul's Hospital, Vancouver, British Columbia[60]	Inpatient	Hip	Male and females Age≥50 years who are likely to return to independent living	NP on the orthopaedic surgery ward refers patients for a bone health assessment via a single page virtual FLS ('vFLS') fax Educational materials provided to patients	N=80 evaluated and recommended appropriate therapy
Lucky Bone™ program Montreal, Quebec[62]	Emergency department and orthopaedic outpatient clinic	Vertebra, sternum, sacrum, wrist, forearm, clavicle, scapula, humerus, ribs, ankle, femur, tibia/fibula, hip and pelvis	Male and females Age≥50 years	Nurse coordinators, identifies patients with fragility fractures and refers to the medical day treatment unit (MDTU) nurses who investigate and initiates pharmacotherapy	190 non-hip fracture patient referred. 97 not seen 54 investigated and treated appropriately 31 refusals 5 deceased, 3 exclusion

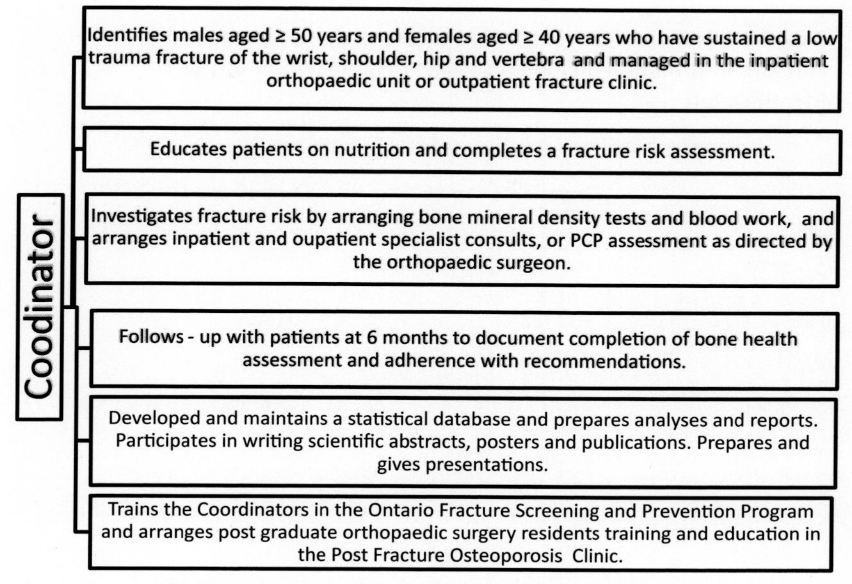

FIGURE 5.1 FLS coordinator role at St. Michael's Hospital.

based on evolving risk assessment tools and treatment guidelines, performance outcomes and qualitative study results. From inception, data collection and analysis have been central to the programme and research ethics board approval was obtained to permit publication of quality assurance data. Between 2002 and 2017 the coordinator was employed through an orthopaedic research fund, with funding obtained through unrestricted research grants from pharmaceutical companies, private donors and OC. Partial funding of the coordinator's salary was provided by the hospital in 2017.

Patient Identification

Men, aged ≥50 years and women aged ≥40 years, with a fracture of the wrist, shoulder, hip or vertebra are identified through a search of the hospital electronic medical records database. The orthopaedic surgeon, resident, fellow or allied health-care professional also alert the coordinator about patients of interest who diverge from the inclusion criteria of the programme. These atypical patients, who are suspected of having bone fragility and elevated risk of refracture, represent approximately 30% of patients managed in this FLS and receive the same bone health assessment as those who meet the inclusion criteria. The total number of patients managed per year ranges from 350 to 450. The coordinator approaches patients who are under the care of an orthopaedic surgeon only. Patients are included in data collection even if they decline intervention, died after identification, have mental, language or physical barriers impacting participation, have contraindications to pharmacotherapy, are not registered for

TABLE 5.3 St. Michael's Hospital FLS Demographic Characteristics and Fracture Information on Patients Not on Pharmacotherapy at Time of Identification

Demographic Characteristics and Fracture Information (N = 2191)		
	Inpatients (n = 862)	**Outpatients (n = 1329)**
Age (year)	78.4 ± 11.9 (40–105 years)	64.9 ± 12.3 (40–99 years)
Fracture Site		
Hip	757 (88%)	59 (4%)
Wrist	31 (4%)	861 (65%)
Shoulder	50 (6%)	389 (29%)
Vertebra	24 (3%)	20 (2%)

public health insurance or reside out of country. In the first 11 years of this programme, from 2002 to 2013, the coordinator identified 4330 patients, of which 3028 met inclusion criteria. Table 5.3 details the demographics and fracture information of patients who met inclusion criteria and were not on pharmacotherapy at the time of identification.

Education

The coordinator educates patients regarding vitamin D and calcium intake and reviews the patient's future fracture risk profile. Educational materials are also provided to the patient and are supplied in hard copy by OC and are also available online at www.osteoporosis.ca. Patients also receive a one-page handout on their future fracture risk and a personalized care pathway (see Fig. 5.2).

Surgeons, residents, fellows and allied health professionals receive continuous education on their role in fracture prevention through presentations at hospital rounds and the sharing of current publications. Orthopaedic residents at the University of Toronto receive training with a multidisciplinary specialist team at the St. Michael's Hospital Post Fracture Osteoporosis (PFO) Clinic which is a formal segment of the orthopaedic residency curriculum. Resident attendance at PFO clinics is arranged by the FLS coordinator in liaison with an administrator at the University of Toronto, Division of Orthopaedic Surgery. In this clinic, restricted to fragility fracture patients referred by the FLS programme, orthopaedic residents learn how to conduct a history and physical exam focused on determining major risk factors for osteoporosis and fragility fractures, the ordering and interpretation of relevant investigations and how to rule out other metabolic bone diseases. They become familiar with the interpretation of FRAX and CAROC for the prediction of the patient's absolute 10-year fracture risk.

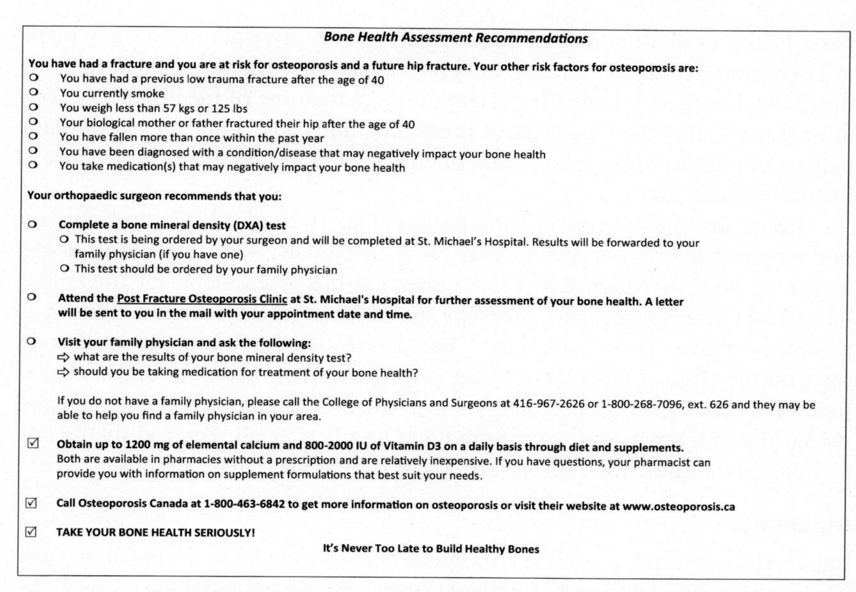

Bone Health Assessment Recommendations

You have had a fracture and you are at risk for osteoporosis and a future hip fracture. Your other risk factors for osteoporosis are:

- o You have had a previous low trauma fracture after the age of 40
- o You currently smoke
- o You weigh less than 57 kgs or 125 lbs
- o Your biological mother or father fractured their hip after the age of 40
- o You have fallen more than once within the past year
- o You have been diagnosed with a condition/disease that may negatively impact your bone health
- o You take medication(s) that may negatively impact your bone health

Your orthopaedic surgeon recommends that you:

o **Complete a bone mineral density (DXA) test**
- o This test is being ordered by your surgeon and will be completed at St. Michael's Hospital. Results will be forwarded to your family physician (if you have one)
- o This test should be ordered by your family physician

o **Attend the Post Fracture Osteoporosis Clinic at St. Michael's Hospital for further assessment of your bone health. A letter will be sent to you in the mail with your appointment date and time.**

o **Visit your family physician and ask the following:**
- ⇨ what are the results of your bone mineral density test?
- ⇨ should you be taking medication for treatment of your bone health?

If you do not have a family physician, please call the College of Physicians and Surgeons at 416-967-2626 or 1-800-268-7096, ext. 626 and they may be able to help you find a family physician in your area.

☑ **Obtain up to 1200 mg of elemental calcium and 800-2000 IU of Vitamin D3 on a daily basis through diet and supplements.** Both are available in pharmacies without a prescription and are relatively inexpensive. If you have questions, your pharmacist can provide you with information on supplement formulations that best suit your needs.

☑ **Call Osteoporosis Canada at 1-800-463-6842 to get more information on osteoporosis or visit their website at www.osteoporosis.ca**

☑ **TAKE YOUR BONE HEALTH SERIOUSLY!**

It's Never Too Late to Build Healthy Bones

FIGURE 5.2 St. Michael's Hospital patient information letter.

They discuss a framework for nonpharmacologic and pharmacologic management with senior staff, recognising important indications and contraindications. Through this clinic training, resident awareness of bone health is enhanced, some residents become independent in managing uncomplicated cases and become efficient in recognising the need for referral for assessment by bone health specialists.

Bone Mineral Density Testing

The coordinator arranges BMD testing under medical directives from the orthopaedic surgeons or requests the patient follow-up with their PCP for a BMD, depending on patient preference.

During the first 11 years of the programme, from 2002 to 2013, of the 2191 patients who were not previously diagnosed and/or on pharmacotherapy at the time of identification, 1599 (73%) were recommended to complete a BMD test and of those 1354 (85%) completed this test.[19]

Low bone mass or osteoporosis was identified by BMD in 91% of patients with a hip fracture, 93% with a vertebral fracture, 77% with shoulder fracture and 68% with a wrist fracture.[19]

Referrals for Bone Health Assessment and Pharmacotherapy

Outpatients

Under direction of the surgeon, the coordinator refers outpatients to a specialist at the Metabolic Bone Clinic (MBC) or PFO Clinic within the hospital, to a specialist at another hospital at the patient's request or to the patient's PCP for bone health evaluation and initiation of treatment, if required. Some patients are managed by the coordinator, under specialist supervision, who were previously diagnosed and treated for osteoporosis at the time of identification, were at moderate rather than high risk of future fracture, had documented contraindications to pharmacotherapy or were unable to attend an assessment because of insurmountable barriers.

During the first 11 years of the programme, 1096 of 1329 outpatients not previously diagnosed and/or treated for osteoporosis were referred to a specialist or PCP of whom 869 (79%) attended the assessment. Of these 869, 453 (52%) received pharmacotherapy, in addition to the recommendation of adequate vitamin D and calcium.[19] The decision to initiate pharmacotherapy rests on the clinical judgement of the physician. There were no differences in treatment rates between men and women. The 233 patients who were managed within the programme or not referred due to a barrier all received the recommendation for guideline appropriate doses of Vitamin D and calcium.

Inpatients

Inpatients are usually assessed by bone health specialists during their stay. Those not assessed during their inpatient stay are referred under the same outpatient pathway or are managed by the coordinator, under specialist supervision, including those with documented contraindications to pharmacotherapy or are unable to attend an outpatient assessment due to insurmountable barriers, such as end of life, terminal cancer and advanced dementia.

A geriatric inpatient assessment programme and a rheumatologist are dedicated to consult on fragility fracture inpatients, and this service was implemented in 2009. Geriatricians assess fragility fracture patients aged ≥65 years and older, and those ≤65 years old are assessed by a rheumatologist. The FLS coordinator ensures that the inpatient team, including the residents, pharmacist, case manager, charge nurse and appropriate specialist are notified of the consultation and that the order is placed in the patient's electronic chart. An electronic inpatient order set provides consistent investigations and vitamin D and calcium supplementation. Consultants communicate the fracture risk and therapeutic recommendations, including pharmacotherapy, to the treating surgeon as well as to the PCP. The coordinator assumes responsibility for follow-up which may include arranging a BMD test and outpatient referrals to a speciality clinic for further assessment.

There were 717 of 862 inpatients, not previously diagnosed and treated for osteoporosis, who were referred to a specialist or PCP and 611 (85%) had a documented assessment. Of these 611 patients, 443 (73%) received pharmacotherapy, in addition to the recommendation of adequate vitamin D and calcium; 87 (14%) received a recommendation for only vitamin D and calcium and 81 (13%) had contraindications to pharmacotherapy and were recommended guideline appropriate doses of vitamin D and calcium.[19] The decision to initiate pharmacotherapy rests on the clinical judgement of the physician. There were no differences in treatment rates between men and women. The 106 patients without a documented bone health assessment and the 145 patients who were managed within the programme or not referred because of barriers, all received a recommendation for guideline appropriate doses of vitamin D and calcium.

Data Collection and Programme Evaluation

During the initial consultation with the patient, the coordinator administers a quality assurance baseline questionnaire completed on a voluntary basis. The data collected includes fracture information, risk factors for fracture, prior BMD testing, currently pharmacotherapy, demographics and knowledge, beliefs and perceptions of fracture and osteoporosis. The 6-month mail-out follow-up questionnaire includes follow-up visits with physicians, BMD testing and new or modified pharmacotherapy and knowledge and awareness of bone health.

The baseline quality assurance questionnaire is not distributed to excessively frail elderly patients, those with profound language, physical or mental barriers to communication or those who decline to participate. The follow-up questionnaire is mailed at 6 months to all patients who had completed a baseline questionnaire and agreed to participate in the follow-up study. Ethics approval is obtained to collect basic baseline and outcomes data through chart audit, on those who did not complete a baseline and/or follow-up questionnaire.

Programme Strengths and Limitations

The strengths of the St. Michael's Hospital FLS are that it is an inclusive, prospective, multidisciplinary and individualized real-world programme. All patients presenting with a fragility fracture or their caregivers receive at least an educational intervention and bone health recommendations. Patients are not excluded from data collection if they decline intervention, die, have mental, language or physical barriers impacting participation or have contraindications to pharmacotherapy.

The coordinator works daily interacting with patients in the Fracture Clinic beside surgeons, residents, fellows and allied health-care professionals. The coordinator also communicates regularly with the inpatient orthopaedic team, Nuclear Medicine technologists, management and clerical staff, specialists and clerical staff at the outpatient MBC and PFO Clinic and with a network of other bone health specialist across the city and province.

The St. Michael's Hospital FLS has produced several publications documenting improved bone health documentation by surgeons,[63] secondary causes of osteoporosis in a fracture population,[64] needs of the wrist fracture population,[65] a cost-effectiveness evaluation,[66] community hospital reproducibility[20] and studies of patients' understanding and behaviours.[67,68]

The St. Michael's Hospital FLS was used as a template for the provincial Ontario Strategy FSPP.[21] The St. Michael's FLS coordinator was seconded to OC to develop the initial protocols for the Ontario programme and continues to train all Ontario coordinators.

This programme was awarded the gold rating of the International Osteoporosis Foundation's Capture the Fracture Initiative and has been recognized by OC for meeting OC's Essential Elements of an FLS.

The limitations of this FLS are that the outcomes, including BMD testing, bone health assessments, and pharmacotherapy are collected through self-report directly to the coordinator and chart audit and are therefore likely conservative. We presume that some additional patients beyond those documented do obtain BMD testing, attend bone health assessments and receive pharmacotherapy, but some confirmatory data are lacking.

COST-EFFECTIVENESS OF FRACTURE LIAISON SERVICES IN CANADA

Two economic studies on FLS programmes in Ontario were conducted[66,69], and several publications have been published on the cost-effectiveness of post-fracture interventions in randomized and nonrandomized controlled trials in Alberta and Manitoba.[70–74] A deterministic cost-effectiveness analysis on the use of a coordinator was completed in the FLS programme at St. Michael's Hospital.[66] According to this analysis an FLS coordinator who manages 500 fragility fracture patients annually can reduce the number of subsequent hip fractures from 34 to 31 in the first year alone, with a net hospital cost savings of $48,950 (Canadian dollars in year 2004 values; $37,000 US dollars in year 2004 values; $30,000 Euros in year 2004 values). Probabilistic sensitivity analysis showed a 90% probability that the cost of hiring a coordinator is less than $25,000 Canadian per hip fracture avoided. Hiring a coordinator is a cost-saving measure even when the coordinator manages as few as 350 patients annually. Greater savings are expected after the first year when further fractures are prevented and additional rehabilitation and dependency costs are considered.

A Markov model was developed to assess the cost-effectiveness of the provincial Ontario FSPP, active at that time in 25 hospital fracture clinics, over the patients' remaining lifetime, using rates of BMD testing and pharmacotherapy and the cost of intervention in the programme.[69] This analysis indicated that quality-adjusted life-years (QALYs) were improved by 4.3 years, with an increased expenditure of $83,000 (Canadian dollars in year 2014 values, $75,000 US dollars in year 2014 values, $56,000 Euros in year 2014 values) for every 1000 patients screened, at a cost of $19,132 per QALY gained (Canadian dollars in year 2014 values, $17,000 US dollars in year 2014 values, $13,000 Euros in year 2014 values). The enhanced BMD Fast Track had greatly superior cost-effectiveness at $5720 per QALY gained (Canadian dollars in year 2014 values, $5000 US dollars in year 2014 values, $3900 Euros in year 2014 values).

The potential impact of systematic uptake of FLS on hip fracture costs in Canada was reported by OC.[75] According to the methodology used for this analysis,[76] if all fracture patients in Canada had access to an FLS and received identification, investigation and initiation of appropriate pharmacotherapy to lower their risk of future fracture, 20,000 hip fractures and 10,000 non-hip fractures would be prevented over an 8-year period. The potential health-care cost saving over that time frame is projected to be over $413 million in acute care hip fracture costs (Canadian dollar in year 2010 values; $400 million US in year 2010 values; $302 million Euros in year 2010 values); additionally, 450,000 acute care hospital bed days would be freed up and 2800 to 4800 long-term care admissions would be averted.

THE PATIENT EXPERIENCE: MAKING FRACTURE LIAISON SERVICES WORK BETTER

A special feature of the Ontario FSPP and the St. Michael's FLS is the unique integration of research on patients' knowledge, perceptions and attitudes, and the use of this information to inform the evolution of these programmes. A team of researchers in Toronto embedded within both the above-named FLS programmes developed a programme of qualitative and quantitative research on patient knowledge and attitudes that guide an iterative process of programme improvement. This approach, partially based on qualitative research, observed that patients at high risk identified by FLS programmes decline to participate or follow through on evidence-based recommendations made in evidence-based hip fracture prevention programmes. A major goal of the research team is to learn about the perspective of the patients on their risk of fracture, the 'alleged' fragility of their bones, their understanding of interventions, their views on the relative risks of using medication or declining to do so and other issues important to effective fracture prevention. Anecdotally, many patients doubt that there is a fragility cause to their fractures, doubt they are at risk for future fracture, are suspicious of interventions and fear side effects from medications. The notion of risk reduction is fundamental to the work of health-care providers, but it is poorly understood among the general population. Education programmes for fragility fracture patients do not always create impactful insights for high fracture risk patients. For these and related reasons, treatment uptake is low in some environments and compliance or persistence with pharmacotherapy is problematic.[17,36] FLS programmes can be effective only if patients engage with interventions and make informed decisions about recommended treatment. For this reason, the Musculoskeletal Health and Outcomes Group at St. Michael's Hospital in Toronto has pursued research to address these and other issues regarding patients' knowledge, beliefs and attitudes around fragility fracture and hip fracture prevention[67,68,77–87]. Some of this work is described below.

Patients Do Not Associate Their Fragility Fracture With Their Underlying Bone Health

A major barrier to treatment initiation (pharmacological and nonpharmacological) is the patient's failure to comprehend the causal connection between their fracture and their personal bone health status.[41,77,78] Beaton and colleagues reported that only 11% of patients associated their fracture with their bone health after having been educated about bone health through the FSPP.[41] Another study revealed that one reason for the 'disconnect' may be miscommunication around the term 'fragility fracture'. In this study, where patients were screened through the St. Michael's Hospital FLS, the term 'fragility fracture' was considered a misnomer by patients who perceived the fracture to be a traumatic experience, both emotionally and physically.[79] The fracture was perceived to be a 'freak' or 'fluke' event and patients used forceful, action-oriented words to describe it.[79] The concept of a fragility fracture, so central to our work in FLS, is resisted and denied by many patients. This and other observations suggest that a radical revision of our knowledge translation approaches and methods may be needed to optimize our communications with fragility fracture patients.

Burden of Fracture

The research team has conducted studies examining the psychological and physical burden of fragility fractures. Few studies have examined fracture-related pain beyond 6 months of fracture. In one study in patients at high risk for future fracture, we found that over two-thirds described fracture-related pain and/or limitations at late follow-up.[80] In a secondary analysis of data sets from three other study samples, we reported that 51% of patients spoke, unprompted, about post-fracture pain experienced beyond 1 year of fracture.[81] Seventy-six percent of these patients had sustained non-vertebral fractures. In many cases, pain was perceived by patients to be disregarded and untreated by health-care providers.

Using qualitative methods, we learned that patients who experience fragility fractures become fearful of falling.[67] In one study, almost all patients noted that they monitored their actions with an unusually elevated degree of vigilance.[67] There is also a degree of work involved in managing one's bone health after a fracture. Patients reported nonpharmacological strategies including exercise, a healthy diet including ingestion of supplements and the use of aids and devices.[67] In another study, we classified individuals as effective consumers based on the intensity of their behaviours that included making requests to health-care providers for referrals to bone specialists, BMD tests, and prescription medication.[82] There was evidence suggesting differences between men and women regarding bone health behaviour after fracture. In one study, men with a wife or female partner described these women as playing an integral role in their bone health, for example by helping with medication regimens and removing tripping hazards.[83] Men also described taking risks such as drinking excessive alcohol and climbing ladders and appeared to not dwell on the meaning of the fracture and/or bone health. These findings suggest that men with fragility fractures rely on their female partners to help them with the work of managing their bone health after a fracture.

Education Is Not Enough

It was demonstrated in several studies that patients screened through the St. Michael's FLS were still unclear about many aspects of their care; they were uncertain about the cause of fracture,[78] the presentation of osteoporosis as a disease,[78] the BMD process and BMD test results[77,78] and treatment recommendations.[78] For example, some patients considered the evidence of compromised bone health was neither accurate nor serious.[77] Many fracture patients remained uncertain about aspects of the management recommended to them after screening such as doses and duration of treatment.[78] These findings support the conclusions of systematic reviews that post-fracture interventions relying solely on education are the least successful in increasing osteoporosis medication initiation rates.[17,36] We are increasingly aware that we are not succeeding in our education goals with a large proportion of patients managed through classic FLS methods, and we wish to discover superior methods of communication/education.

Pharmacotherapy Use: Uptake and Persistence

Approximately one-half of patients have difficulty making the decision to take pharmacotherapy to lower the risk of future fracture because of concerns about perceived as well as proven side effects of the medication.[84] We found that osteoporosis medication use may vary over time as patients may start, stop, and then restart medication.[84] Individuals 65 years and older screened through the St. Michael's Hospital FLS who were at high risk for future fracture indicated that they might be persuaded to start or stop pharmacotherapy depending on a number of factors.[84] In another study, 13% of patients who were on pharmacotherapy at baseline had discontinued therapy within 6 months of screening.[20] Our findings suggest that health-care providers should periodically 'check in' with patients to determine if willingness to take pharmacotherapy has changed.[84]

Reliance on the Concept of 'Fracture Risk' May Be Problematic

One reason that fracture patients who had indications for pharmacotherapy but did not initiate treatment may be the failure of the patient to learn their level of future fracture risk through BMD testing and to absorb the meaning of the risk. In one study of patients screened through the FSPP, it was found that BMD test reports underestimated fracture risk because prevalent or prior fractures were disregarded as risk factors in the fracture risk assessment.[85] The concept of fracture risk is difficult for some patients to understand. An intervention was conducted where fracture risk was communicated to patients at high risk for future fracture and to their health-care providers using three communication strategies (verbal, numerical, visual graph).[68] Patients overestimated the probability of re-fracture and had poor recall of the visual graph. Of those who recalled being deemed to be at 'high risk', approximately one-half did not believe that it applied to them. Some patients were unsure of what they were at risk for, citing being at risk for osteoporosis, low bone density and/or falling.

Barriers at the Health-Care Provider Level

PCPs may be partly responsible for gaps in post-fracture care. An analysis was conducted of what patients report about interactions with PCP after fragility fractures. Some physicians use incorrect heuristics to make clinical judgements, such as telling patients that they do not require BMD testing because they are overweight, too young, too old, or from a specific ethnic group.[86] Patients also report that, when compared to their bone health specialists, patients perceived their PCPs as being uninterested in their bone health despite patients presenting with height loss, vertebral fractures, a history of multiple fractures, a family history of osteoporosis, or being otherwise at high risk for future fracture.[87] Participants in this study also perceived inconsistent messages about bone health and bone health management recommendations within, and across, health care providers.

CONCLUSIONS

The federal political structure of Canada with decentralization of health-care management and finance to the provinces and territories is a barrier to the establishment of a universal fracture prevention system. OC, the national non-profit organisation, provides linkage, documentation and support across the country. Three provinces, Ontario, Alberta and Nova Scotia, have provincially funded FLS. Separate initiatives have established FLS programmes in British Columbia, Ontario and Quebec.

In provinces and territories where system-wide FLS has not been initiated, the challenge is to educate and influence politicians and senior Health Ministry bureaucrats to institute FLS programmes. Our experiences to date in three provinces indicate strongly that cost-effectiveness data, increasingly available from high quality studies, are key motivators for decision-makers.

In areas where system level support has not been achieved, it is essential to identify a local health manager or hospital executive willing to champion the FLS cause—a challenge in a health-care administrative structure where osteoporosis has no home. It requires months or years of effort for the FLS champion, often an orthopaedic surgeon or bone health specialist, and the forward-thinking hospital executive, to win support for financing of an FLS programme.

Optimists viewing the increasing penetration of FLS in Canada consider that we are nearing the 'tipping point' where it will be considered mandatory to provide system-wide fracture prevention, and health ministries will require hospital performance in this area in their funding models.

There are notable positive features evident in some Canadian FLS programmes, beyond the essential elements shared by FLS programmes in general. The Canadian FLS experience has permitted us to see the importance of:

1. Iterative improvement processes
   Programmes that self-assess for intermediate outcomes and repeatedly cycle through improvement initiatives can rapidly evolve to overcome local

barriers and other difficulties. The Canadian experience has shown that an open process of cyclical improvement based on identifying barriers and failures and introducing innovations results in documented programme improvement.

2. Adaptability to local conditions
The successful delivery of fracture prevention interventions cannot be homogeneous in form across all health-care environments. Local conditions affect the performance and ideal design of FLS programmes. For example, the pattern of concentration of fragility fracture patients determines the locus of efficient and effective screening. In Canada, this locus is usually the hospital fracture clinic and orthopaedic inpatient ward where intensive work can be profitably focused. Health-care funding models influence the opportunities and challenges facing FLS programmes. The local availability and funding of BMD and access to pharmacotherapy are important.
The Canadian experience has demonstrated that FLS programmes benefit from sensitivity to local conditions. For example, rural areas which are sparsely populated and medically underserviced require different models of FLS and may benefit from a regional coordinator who accesses patients by telephone. A 'virtual FLS' has also been innovated in an urban centre where FLS funding has not yet been secured.

3. The importance of the coordinator/FLS nurse role
The Canadian FLS experience confirms the now widely accepted principle that the FLS nurse/coordinator has the key role in the successful provision of FLSs. It is this individual, supported by the FLS champion and the lead orthopaedic surgeon (who may be the same person), who ensures high-quality care for each fragility fracture patient arriving in the orthopaedic/fracture clinic or inpatient ward. This individual is also the one who first encounters obstacles and difficulties and must adapt the processes of the FLS to successfully deal with local conditions.

4. Focused data collection to enable process evaluation and outcome assessment
This is a feature of FLS which is necessary to monitor performance and enable programme improvement as difficulties and failures are encountered. To facilitate the analysis and improvement of FLS programmes, it is advantageous to build into the design a robust searchable data collection and storage system. This process of data collection and banking should itself be the object of cycles of improvement based on performance. Some Canadian programmes built the collection of data regarding programme performance and patient knowledge and perceptions into the FLS from inception.

5. Gender equality was achieved through two Canadian FLS Programmes
Males at high risk of future fracture received equal rates of investigation and treatment when compared with females of the same.[19,58]

6. The FLS as an engine for education
The FLS at St. Michael's Hospital leverages the activities of the FLS coordinator and of the bone health specialists to provide education on

fracture prevention to orthopaedic residents in training. Every effort is made to increase awareness of fracture prevention in nurses, physical therapists, occupational therapists, orthopaedic technologists, medical and nursing students as well as patients and their relatives and caregivers.

7. The importance of the patient's knowledge, perceptions and attitudes
Pioneering investigators in Canada documented what many suspect: the patient perceptions of their fracture risk, their bone fragility and the benefits and disadvantages of fracture prevention interventions differ greatly from the perceptions of health-care providers. Lack of agreement on these issues is probably a principal cause of failure of patients at high fracture risk to engage in hip fracture interventions and of poor adherence with pharmacotherapy. We believe that a thorough review of the assumptions and methods which govern our communications with patients is required. This is what we consider to be the 'next frontier' in FLSs.

THE FUTURE OF FRACTURE LIAISON SERVICES IN CANADA

The key challenge for improvement of fracture prevention in Canada is to advocate for and establish high-functioning FLS programmes throughout all 13 health-care jurisdictions (the provinces and territories). Effective programmes currently operate in three provinces. Efforts are underway by OC, the national nonprofit organisation, and by individual groups, to expand the programmes across the country.

The existence of large provincial health-care databases in each jurisdiction should facilitate the growth of FLS and related fracture prevention programmes, if privacy concerns can be properly accommodated and adequate governmental funding secured.

ACKNOWLEDGEMENTS

Sonia Singh, MD. Regional Medical Director, Research and Evaluation Fraser Health Peace Arch Hospital, White Rock, BC, Canada.

David Kendler, MD. Professor of Medicine, University of British Columbia, Vancouver, BC, Canada.

Diane Theriault, MD. *Osteoporosis Canada's Chief Scientific Officer of FLS*. Dartmouth, NS, Canada.

Alexandra Papaioannou, MD. Executive Director, GERAS Centre; Geriatrician, Hamilton Health Sciences; Professor of Medicine, Department of Medicine, McMaster University, Hamilton, ON, Canada.

Aliya Khan, MD. Professor of Clinical Medicine, Divisions Endocrinology and Metabolism and Geriatrics, McMaster University, Hamilton, ON, Canada.

Amanda Pellecchia, MPH. Quality Improvement Manager - Fragility & Stability Alberta Bone & Joint Health Institute, Calgary, AB, Canada.

Angela M. Cheung, MD, PhD. Staff General Internist. Director, Osteoporosis Program; Director, Centre of Excellence in Skeletal Health Assessment; Director, Women's Health Program at University Health Network/Toronto Rehabilitation Institute/Mount Sinai Hospital. Professor of Medicine, University of Toronto, Toronto, ON, Canada.

REFERENCES

1. Canada. Statistics Canada. *Population by year, by province and territory*. 2017. http://www.statcan.gc.ca/tables-tableaux/sum-som/l01/cst01/demo02a-eng.htm.
2. Canada. Statistics Canada. *Population size and growth in Canada: key results from the 2016 Census*. 2017. http://www.statcan.gc.ca/daily-quotidien/170208/dq170208a-eng.htm.
3. Canada. Statistics Canada. *Population by sex and age group*. 2017. http://www.statcan.gc.ca/tables-tableaux/sum-som/l01/cst01/demo10a-eng.htm.
4. Canada. Statistics Canada. *Life expectancy*. 2015. https://www.statcan.gc.ca/pub/89-645-x/2010001/life-expectancy-esperance-vie-eng.htm.
5. Canada. Statistics Canada. *Population projections for Canada. Section 2 – Results at the Canada level, 2013 to 2063*. 2015. https://www.statcan.gc.ca/pub/91-520-x/2014001/section02-eng.htm.
6. Prior JC, Langsetmo L, Lentle BC, et al. Ten-year incident osteoporosis-related fractures in the population-based Canadian Multicentre Osteoporosis Study - comparing site and age-specific risks in women and men. Research Group. *Bone* 2015;**71**:237–43.
7. Hopkins RB, Burke N, Von Keyserlingk C, et al. The current economic burden of illness of osteoporosis in Canada. *Osteoporos Int* 2016;**27**(10):3023–32.
8. Klotzbuecher CM, Ross PD, Landsman PB, Abbott 3rd TA, Berger M. Patients with prior fractures have an increased risk of future fractures: a summary of the literature and statistical synthesis. *J Bone Miner Res* 2000;**15**(4):721–39.
9. Edwards BJ, Bunta AD, Simonelli C, Bolander M, Fitzpatrick LA. Prior fractures are common in patients with subsequent proximal femur fractures. *Clin Orthop Relat Res* 2007;**461**:226–30.
10. Giangregorio LM, Leslie WD. Manitoba Bone Density Program. Time since prior fracture is a risk modifier for 10-year osteoporotic fractures. *J Bone Miner Res* 2010;**25**(6):1400–5.
11. Ioannidis G, Papaioannou A, Hopman WM, et al. Relation between fractures and mortality: results from the Canadian Multicentre Osteoporosis Study. *CMAJ* 2009;**81**(5):265–71.
12. Morin S, Lix LM, Azimaee M, Metge C, Caetano P, Leslie WD. Mortality rates after incident non-traumatic fractures in older men and women. *Osteoporos Int* 2011;**22**(9):2439–48.
13. Beaupre LA, Jones CA, Johnston DW, Wilson DM, Majumdar SR. Recovery of function following a proximal femur fracture in geriatric ambulatory persons living in nursing homes: prospective cohort study. *J Am Geriatr Soc* 2012;**60**(7):1268–73.
14. Adachi JD, Loannidis G, Berger C, et al. The influence of osteoporotic fractures on health-related quality of life in community-dwelling men and women across Canada. *Osteoporos Int* 2001;**12**(11):903–8.
15. Tarride JE, Burke N, Leslie WD, et al. Loss of health-related quality of life following low-trauma fractures in the elderly. *BMC Geriatr* 2016;**19**(16):84.
16. Kerr C, Bottomley C, Shingler S, et al. The importance of physical function to people with osteoporosis. *Osteoporos Int* 2017;**28**(5):1597–607.
17. Sale JE, Beaton D, Posen J, Elliot-Gibson V, Bogoch E. Systematic review on interventions to improve osteoporosis investigation and treatment in fragility fracture patients. *Osteoporos Int* 2011;**22**(7):2067–82.
18. Bogoch ER, Elliot-Gibson V, Beaton DE, Jamal SA, Josse RG, Murray TM. Effective initiation of osteoporosis diagnosis and treatment for patients with a fragility fracture in an orthopaedic environment. *J Bone Joint Surg Am* 2006;**88**(1):25–34.
19. Bogoch ER, Elliot-Gibson V, Beaton D, Sale J, Josse RG. Fracture prevention in the orthopaedic environment: outcomes of a coordinator-based fracture liaison service. *J Bone Joint Surg Am* 2017;**99**(10):820–31.

20. Sale JE, Beaton DE, Elliot-Gibson VI, Bogoch ER, Ingram J. A post-fracture initiative to improve osteoporosis management in a community hospital in Ontario. *J Bone Joint Surg Am* August 18, 2010;**92**(10):1973–80.
21. Jaglal SB, Hawker G, Cameron C, et al. The Ontario Osteoporosis Strategy: implementation of a population-based osteoporosis action plan in Canada. *Osteoporos Int* 2010;**21**(6):903–8.
22. Osteoporosis Canada. *Canadian FLS registry map.* 2017. http://fls.osteoporosis.ca/canadian-fls-registry/.
23. Roux S, Beaulieu M, Beaulieu MC, Cabana F, Boire G. Priming primary care physicians to treat osteoporosis after a fragility fracture: an integrated multidisciplinary approach. *J Rheumatol* 2013;**40**(5):703–11.
24. Gaboury I, Corriveau H, Boire G, Cabana F, Beaulieu MC, Dagenais P, et al. Partnership for fragility bone fracture care provision and prevention program (P4Bones): study protocol for a secondary fracture prevention pragmatic controlled trial. *Implement Sci* 2013;**4**(8):10.
25. Canada. Government of Canada. *Canada's health care system.* 2017. https://www.canada.ca/en/health-canada/services/canada-health-care-system.html.
26. Papaioannou A, Morin S, Cheung AM, et al. 2010 clinical practice guidelines for the diagnosis and management of osteoporosis in Canada: summary. *CMAJ* 2010;**182**(17):1864–73.
27. Ontario. Ministry of Health and Long-Term Care, Government of Ontario. *Bone mineral density (BMD) testing.* 2013. http://www.health.gov.on.ca/en/public/publications/ohip/bone.aspx.
28. Nova Scotia. Government of Nova Scotia. *Bone density test.* 2013. https://811.novascotia.ca/health_topics/bone-density-test/.
29. Alberta Medical Association Fee Navigator. *Health service code X128.* 2017. https://www.albertadoctors.org/fee-navigator/hsc/X128.
30. *List of medications for osteoporosis available in Canada.* Osteoporosis Canada; 2018. https://osteoporosis.ca/about-the-disease/treatment/provincial-drug-coverage/.
31. Leslie WD, Lix LM, Langsetmo L, et al. Construction of a FRAX® model for the assessment of fracture probability in Canada and implications for treatment. *Osteoporos Int* 2011;**22**(3):817–27.
32. Leslie WD, Lix LM, Johansson H, Oden A, McCloskey E, Kanis JA. Manitoba Bone Density Program. Independent clinical validation of a Canadian FRAX tool: fracture prediction and model calibration. *J Bone Miner Res* 2010;**25**(11):2350–8.
33. Kanis JA, Harvey NC, Cooper C, Johansson H, Odén A, McCloskey EV. Advisory board of the national osteoporosis guideline group. A systematic review of intervention thresholds based on FRAX : a report prepared for the national osteoporosis guideline group and the international osteoporosis foundation. *Arch Osteoporos* 2016;**11**(1):25.
34. Osteoporosis Canada. *Osteoporosis Canada's role in supporting the implementation of quality FLSs.* December, 2017. http://fls.osteoporosis.ca/wp-content/uploads/oc-role-implementation-canadian-fls.pdf. Reprinted with permission.
35. Osteoporosis Canada. *Osteoporosis: towards a fracture free future.* 2011. http://osteoporosis.ca/fls/wp-content/uploads/white-paper-march-2011.pdf. Toronto, Ont.
36. Ganda KPM, Chen JS, Speerin R, Beasel J, Center JR, Eisman JA, et al. Models of care for the secondary prevention of osteoporotic fractures: a systematic review and meta-analysis. *Osteoporos Int* 2013;**24**:393–406.
37. *Healthcare tomorrow.* 2017. http://healthcaretomorrow.ca/wp-content/uploads/2015/03/Understanding-the-Health-Care-System.pdf.
38. *Osteoporosis action plan: an osteoporosis strategy for Ontario report of the osteoporosis action plan committee to the Ministry of Health and Long-Term Care.* February 2003. http://www.ontla.on.ca/library/repository/mon/10000/250502.pdf.

39. *Ontario osteoporosis strategy*. 2017. http://www.osteostrategy.on.ca/.
40. Beaton DE, Sujic R, McIlroy BK, Sale J, Elliot-Gibson V, Bogoch ER. Patient perceptions of the path to osteoporosis care following a fragility fracture. *Qual Health Res* 2012;**22**:1647–58.
41. Beaton DE, Dyer S, Jiang D, Sale JEM, Slater M, Bogoch E. Factors influencing the pharmacological management of osteoporosis after a fragility fracture: results from the Ontario osteoporosis strategy's fracture clinic screening program. *Osteoporos Int* 2014;**25**:289–96.
42. Sujic R, Gignac MA, Cockerill R, Beaton DE. Factors predictive of the perceived osteoporosis-fracture link in fragility fracture patients. *Maturitas* 2013;**76**:179–84.
43. Beaton DE, Vidmar M, Pitzul KB, Sujic R, Rotondi NK, Bogoch ER, et al. Addition of a fracture risk assessment to a coordinator's role improved treatment rates within 6 months of screening in a fragility fracture screening program. *Osteoporos Int* 2017;**28**:863–9.
44. Jain R. Osteoporosis Canada. *Email communication*. Email date: November 27, 2017.
45. Ansari H, Beaton DE, Sujic R, Rotondi NK, Cullen JD, Slater M, et al. Ontario Osteoporosis Strategy Fracture Screening and Prevention Program Evaluation Team Equal treatment: no evidence of gender inequity in osteoporosis management in a coordinator-based fragility fracture screening program. *Osteoporos Int* 2017;**28**(12):3401–6.
46. Masud T, Jordan D, Hosking DJ. Distal forearm fracture history in an older community-dwelling population: the Nottingham Community Osteoporosis (NOCOS) study. *Age Ageing* 2001;**30**:255–8.
47. Papaioannou A, Kennedy CC, Ioannidis G, Gao Y, Sawka AM, Goltzman D, et al. CaMos research group. The osteoporosis care gap in men with fragility fractures: the Canadian multicentre osteoporosis study. *Osteoporos Int* 2008;**19**(4):581–7.
48. McLaughlin, Kaluzny. *Continous quality improvement in health care: theory: implementation and applications*. 2nd ed. 2004.
49. Jain R. Osteoporosis Canada. *Email communication*. November 17, 2017.
50. Nova Scotia Department of Health and Wellness. *Acute and chronic disease target setting project – summary report*. April 2012.
51. Theriault D, Purcell C. A Fracture Liaison Service specifically designed to address local government concerns can be effective. *J Bone Miner Res* 2014;**28**(Suppl. 1). Available at: http://www.asbmr.org/education/AbstractDetail?aid=c0d63784-f815-462c-91fc-50f63ded85d1.
52. Osteoporosis Canada. *FLS indicators*. 2017. http://www.osteoporosis.ca/fls/indicator.
53. Amanda Pellecchia. Personal Communication. January 3, 2018.
54. *Hamilton health sciences. Fracture?… Think osteoporosis*. 2017. https://fhs.mcmaster.ca/medicine/geriatric/docs/Fracture-Alert-Newsletter-1-April-2009-FINAL.pdf.
55. Dore N, Kennedy C, Fisher P, Dolovich L, Farrauto, Papaioannou A. Improving care after proximal femur fracture: the fracture? Think Osteoporosis (FTOP) Program. *BMC Geriatr* 2013;**13**:130.
56. Cheung A. *Email communication*. December 18, 2017.
57. Khan A. *Email communication*. November 23, 2017.
58. British Columbia. Government of British Columbia. 2017. https://www2.gov.bc.ca/gov/content/health/practitioner-professional-resources/bc-guidelines/osteoporosis.
59. Singh S. Breaking the cycle of recurrent fracture: implementation of a fracture liaison service in British Columbia, Canada. In: *Presented at the fragility fracture network meeting, Rome, Sept 1–3*. Fragility Fracture Network; 2016.
60. Kendler D. *Email communication*. December 14, 2017.
61. Quebec. Government of Quebec. 2013. http://publications.msss.gouv.qc.ca/msss/fichiers/2012/12-272-01W.pdf.

62. Senay A, Delisle J, Giroux M, Laflamme GY, Leduc S, Malo M, Nguyen H, Ranger P, Fernandes JC. The impact of a standardized order set for the management of non-hip fragility fractures in a Fracture Liaison Service. *Osteoporos Int* 2016;**27**(12):3439–47.
63. Ward SE, Laughren JJ, Escott BG, Elliot-Gibson V, Bogoch ER, Beaton DE. A program with a dedicated coordinator improved chart documentation of osteoporosis after fragility fracture. *Osteoporos Int* 2007;**18**(8):1127–36.
64. Bogoch ER, Elliot-Gibson V, Wang RY, Josse RG. Secondary causes of osteoporosis in fracture patients. *J Orthop Trauma* 2012;**26**(9):e145–52.
65. Bogoch ER, Elliot-Gibson V, Escott BG, Beaton DE. The osteoporosis needs of patients with distal radius fracture. *J Orthop Trauma* 2008;**22(8)**(Suppl.):S73–8.
66. Sander B, Elliot-Gibson V, Beaton DE, Bogoch ER, Maetzel A. A coordinator program in post-fracture osteoporosis management improves outcomes and saves costs. *J Bone Joint Surg Am* 2008;**90**(6):1197–205.
67. Sale JE, Gignac MA, Hawker G, Beaton D, Bogoch E, Webster F, Frankel L, Elliot-Gibson V. Non-pharmacological strategies used by patients at high risk for future fracture to manage fracture risk—a qualitative study. *Osteoporos Int* 2014;**25**(1):281–8.
68. Sale JE, Gignac MA, Hawker G, Beaton D, Frankel L, Bogoch E, Elliot-Gibson V. Patients do not have a consistent understanding of high risk for future fracture: a qualitative study of patients from a post-fracture secondary prevention program. *Osteoporos Int* 2016;**27**(1):65–73.
69. Yong JH, Masucci L, Hoch JS, Sujic R, Beaton D. Cost-effectiveness of a fracture liaison service–a real-world evaluation after 6 years of service provision. *Osteoporos Int* 2016;**27**(1):231–40.
70. Majumdar SR, Johnson JA, Lier DA, Russell AS, Hanley DA, Blitz S, Steiner IP, Maksymowych WP, Morrish DW, Holroyd BR, Rowe BH. Persistence, reproducibility, and cost-effectiveness of an intervention to improve the quality of osteoporosis care after a fracture of the wrist: results of a controlled trial. *Osteoporos Int* March 2007;**18**(3):261–70.
71. Majumdar SR, Lier DA, Beaupre LA, Hanley DA, Maksymowych WP, Juby AG, Bell NR, Morrish DW. Osteoporosis case manager for patients with proximal femur fractures: results of a cost-effectiveness analysis conducted alongside a randomized trial. *Arch Intern Med* 2009;**169**(1):25–31.
72. Majumdar SR, Johnson JA, Bellerose D, McAlister FA, Russell AS, Hanley DA, Garg S, Lier DA, Maksymowych WP, Morrish DW, Rowe BH. Nurse case-manager vs multifaceted intervention to improve quality of osteoporosis care after distal radius fracture: randomized controlled pilot study. *Osteoporos Int* 2011;**22**(1):223–30.
73. Majumdar SR, Lier DA, Rowe BH, Russell AS, McAlister FA, Maksymowych WP, Hanley DA, Morrish DW, Johnson JA. Cost-effectiveness of a multifaceted intervention to improve quality of osteoporosis care after distal radius fracture. *Osteoporos Int* 2011;**22**(6):1799–808.
74. Majumdar SR, Lier DA, Leslie WD. Cost-effectiveness of two inexpensive post fracture osteoporosis interventions: results of a randomized trial. *J Clin Endocrinol Metab* 2013;**98**(5):1991–2000.
75. http://fls.osteoporosis.ca/wp-content/uploads/FLS-TOOLKIT-App-F.pdf.
76. McLellan AR, Wolowacz SE, Zimovetz EA, et al. Fracture liaison services for the evaluation and management of patients with osteoporotic fracture: a cost-effectiveness evaluation based on data collected over 8 years of service provision. *Osteoporos Int* 2011;**22**:2083–98.
77. Sale JEM, Beaton D, Fraenkel L, Elliot-Gibson V, Bogoch E. The BMD muddle: the disconnect between bone densitometry results and perception of bone health. *J Clin Densitom* 2010;**13**(4):370–8.
78. Sale JEM, Beaton DE, Sujic R, Bogoch ER. "If it was osteoporosis, I would have really hurt myself". Ambiguity about osteoporosis and osteoporosis care despite a screening program to educate fracture patients. *J Eval Clin Pract* 2010;**16**(3):590–6.

79. Sale JEM, Gignac M, Hawker G, Frankel L, Beaton D, Bogoch E, et al. Patients reject the concept of fragility fracture - a new understanding based on fracture patients' communication. *Osteoporos Int* 2012;**23**(12):2829–34.
80. Sale JEM, Frankel L, Thielke S, Funnell L. Pain and fracture-related limitations persist 6 months after a fragility fracture. *Rheumatol Int* 2017;**37**:1317–22.
81. Gheorghita A, Webster F, Thielke S, Sale JEM. *Long-term experiences of pain after a fragility fracture. Osteoporos Int* 2018;**29**(5):1093–1104.
82. Sale JEM, Cameron C, Hawker G, Jaglal S, Funnell L, Jain R, et al. Strategies use by an osteoporosis patient group to navigate for bone health care after a fracture. *Arch Orthop Trauma Surg* 2014;**134**:229–35.
83. Sale JEM, Ashe MC, Beaton D, Bogoch E, Frankel L. Men's health-seeking behaviours regarding bone health after a fragility fracture: a secondary analysis of qualitative data. *Osteoporos Int* 2016;**27**(10):3113–9.
84. Sale J, Gignac M, Hawker G, Frankel L, Beaton D, Bogoch E, et al. Decision to take osteoporosis medication in patients who have had a fracture and are 'high' risk for future fracture. *BMC Musculoskelet Disord* 2011;**12**:92.
85. Sale JEM, Bogoch E, Meadows L, Gignac M, Frankel L, Inrig T, et al. Bone mineral density reporting underestimates fracture risk in Ontario. *Health* 2015;**7**:566–71.
86. Sale JEM, Bogoch E, Hawker G, Gignac M, Beaton D, Jaglal S, et al. Patient perceptions of provider barriers to post-fracture secondary prevention. *Osteoporos Int* 2014;**25**:2581–9.
87. Sale JEM, Hawker G, Cameron C, Bogoch E, Jain R, Beaton D, et al. Perceived messages about bone health after a fracture are not consistent across health care providers. *Rheumatol Int* 2015;**35**(1):97–103.

Chapter 6

The Challenge of Secondary Prevention of Hip Fracture in Japan

Noriaki Yamamoto[1], Hideaki E. Takahashi[1], Naoto Endo[2]
[1]*Department of Orthopedic Surgery, Niigata Rehabilitation Hospital, Niigata, Japan;*
[2]*Department of Orthopedic Surgery, Niigata University School of medicine, Niigata, Japan*

BACKGROUND

Japan is one of the most rapidly aging countries in the world, with the population of elderly people over 65 years exceeding 28%. The estimated hip fracture incidence is over 200,000 annually, and still increasing every year. Patients who have suffered an osteoporotic fracture are at high risk of future fracture, but many patients do not receive treatment for their osteoporosis after a sentinel fracture. Indeed, it has been reported that on occasion, osteoporosis therapy has been stopped for hip fracture patients.[1]

This chapter introduces the ongoing activities related to secondary fracture prevention in Japan.

HIP FRACTURE CLINICAL PATHWAY IN JAPAN

In 2006, the Ministry of Health, Labor and Welfare launched the Regional Cooperation of Clinical Pathways for Hip Fracture in Japan. This was the first regional pathway for this particular condition with additional medical fees, so the system was spread to the whole of Japan. However, the aim of this clinical pathway was mainly focused on shortening the length of stay in acute hospitals, whereas treatment rate for osteoporosis remained very low in both acute and recovery hospitals. In 2012, the Japanese Orthopaedic Association (JOA) reported that osteoporosis treatment rates in the hip fracture clinical pathway for the whole of Japan were only 9.8% in acute hospitals and 17.8% in recovery hospitals.[2]

The reasons why fracture patients do not receive treatment for their osteoporosis is a consequence of several reimbursement mechanisms within the Japanese health-care system, such as per diem and bundled payment systems, nursing care insurance and other administrative obstacles.

Secondary Fracture Prevention. https://doi.org/10.1016/B978-0-12-813136-7.00006-5

OSTEOPOROSIS LIAISON SERVICE AND OSTEOPOROSIS MANAGER

In 2015, the Japan Osteoporosis Society (JOS) initiated an osteoporosis manager certification system for the osteoporosis liaison service (OLS) model of care. To maximise the impact of preventive efforts, on account of the huge burden facing Japan in the coming decades, in addition to delivering secondary fracture prevention through the Fracture Liaison Service (FLS) model of care, the OLS model includes efforts to identify and manage individuals who are at high risk of sustaining a first fragility fracture. The osteoporosis managers are certified pursuant to a program of educational lectures and an examination. As of March 2017, a total of 1867 managers had been trained and certified. These osteoporosis managers are expected to coordinate the multidisciplinary approach to establish a FLS, but at the time of writing, only a very small proportion of Japanese hospitals have done this for hip fracture patients, and fewer still for patients presenting with fragility fractures at other relevant skeletal sites (e.g., wrist, humerus, spine and pelvis).

FRAGILITY FRACTURE NETWORK JAPAN

The fragility fracture network (FFN) is a global organisation, which was founded to create a multidisciplinary network of experts for improving treatment and secondary prevention of fragility fractures. The FFN's vision and mission underline this and are the foundation for all decisions taken (http://fragilityfracturenetwork.org/our-organisation/ffn-strategic-plan-2017-2021/). The FFN believes that useful policy change can only happen at a national level, and multidisciplinary national coalitions are the most effective way to achieve this. Hence, the FFN acts as a global template for creating national alliances in as many countries as possible.

The FFN embraces all relevant disciplines including orthopaedics, geriatrics/internal medicine, rehabilitation physicians, radiologists, anaesthetics, endocrinologists, rheumatologists, nursing, physiotherapy and other allied health professionals and relevant researchers. The two main cornerstones are as follows:

- An orthogeriatric approach to the acute fracture episode in older people, who often present with complex medical problems and frailty, which require a geriatric/internist approach.
- FLSs for secondary prevention in fragility fracture patients of all ages.

The network should be composed of activists who, in their home countries, work through their own professional organisations as catalysts and change agents. FFN should be strong enough to raise awareness, achieve policy change and ultimately improve treatment for patients with fragility fracture. National FFNs have been established in China, Greece, India, Japan and Norway, and discussions to establish more national FFNs are ongoing in many other countries.

	Chairman & place	agenda	Participants
2012	• 1st : Hiroshi Hagino • Kyoto	11 lectures	130 Physician 48
2104	• 2nd : Naoto Endo • Niigata	9 lectures 1 workshop	160 Physician 62
2015	• 3rd : Takeshi Sawaguchi • Toyama	11 lectures, symposium 10 free papers	163 Physician 81
2016	• 4th : Masahiro Shirahama • Hukuoka	16 lectures 18 free papers	165 Physician 99
2017	• 5th : Noriaki Yamamoto • Niigata	26 lectures , symposium 30 free papers	255 Physician 90
2018	• 6th : Satoshi Mori • Hamamatsu	28 lectures 29 free papers	250 Physician 71

FIGURE 6.1 FFN Japan annual meeting.

In 2012, the first FFN Japan (FFN-J) meeting was held in Kyoto. Since then, annual meetings of FFN-J have been convened with the number of participants exceeding 300 at the fifth FFN-J meeting in 2018 (Fig. 6.1). In 2015, FFN-J (President Takashi Matsushita) was formally established as a nonprofit organisation and created a website to support new activities (http://ffn.or.jp/). The aim of FFN-J is to achieve the best multidisciplinary care in managing fragility fractures in Japan with connection to the global FFN organisation. FFN-J is in the process of developing a collaboration with the JOA, JOS, Japanese Society for Fracture Repair, Japan Geriatrics Society and Japanese Society for Clinical Pathway. During the fifth FFN-J meeting held in March 2017, a round table symposium on best practice in hip fracture care included a discussion on such a collaborative approach by all relevant societies. In March 2018, Dr. Masaki Fujita from the International Osteoporosis Foundation (IOF) was invited to present on the IOF Capture the Fracture programme at the sixth FFN-J meeting. Industry partners from the pharmaceutical sector have provided ongoing support to these educational initiatives.

NATIONAL HIP FRACTURE DATABASE JAPAN PROJECT

In 2017, FFN-J announced a project to develop a Japan National Hip Fracture Database (JNHFD). The FFN minimum common dataset for hip fracture was translated into Japanese. The ethics committee of Niigata University and Fukushima Medical University approved this project, and 20 hospitals joined

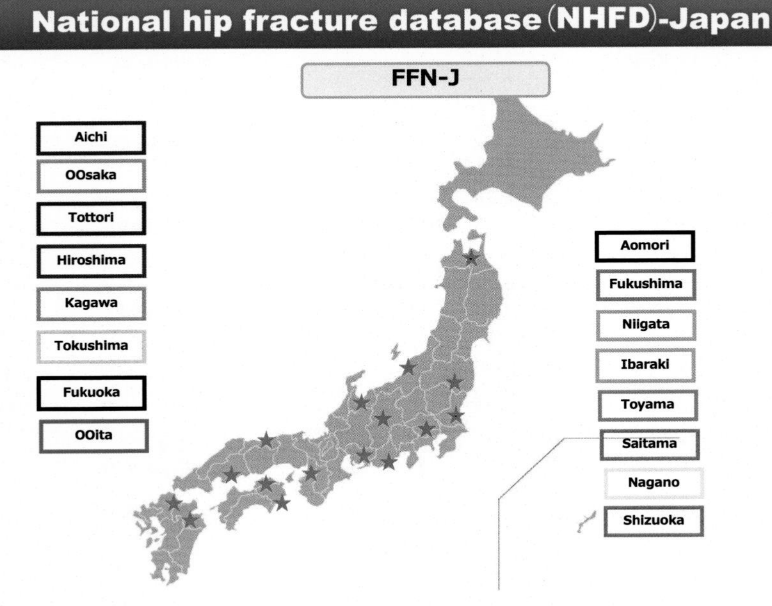

FIGURE 6.2 FFN Japan National Hip Fracture Database Project in Japan.

this project during 2017, with data entry commencing in August 2017 (Fig. 6.2). FFN-J intends to drive nationwide roll-out of the JNHFD and to develop a financial incentive scheme which links quality of care to reimbursement, akin to the best practice tariff developed in the United Kingdom which is discussed in Chapter 10. These initiatives are intended to change health-care policy for fragility fracture management in Japan.

Bone and Joint Japan has begun to promote secondary fracture prevention to all medical staff and lay people. A poster which describes the awareness of secondary fracture risk in the elderly has been distributed throughout to Japan, with the catchphrase 'Stop the fracture domino' (Fig. 6.3).

PROJECTS IN NIIGATA PREFECTURE

The Department of Orthopedic Surgery at Niigata University conducted the first hip fracture survey in Japan in 1985.[3] At the time, the crude hip fracture incidence rate was only 132.1 per 100,000 of population per year, which had increased dramatically to 385.0 per 100,000 by 2010.[4] In 2012, the multidisciplinary team for secondary prevention service was established in Niigata Rehabilitation Hospital. The One-Stop Notebook was developed, which contained all information concerning prevention of secondary hip fractures and to clarify the purpose and role of each staff member in the FLS team (Fig. 6.4). A multidisciplinary

FIGURE 6.3 Bone and Joint Japan poster of secondary fracture prevention.

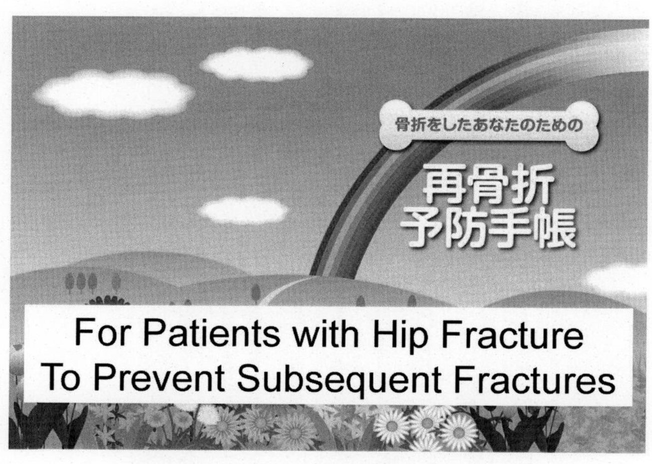

FIGURE 6.4 One-Stop Notebook in Niigata.

TABLE 6.1 Trend of the Incidence of Hip Fractures in Niigata Prefecture

Survey Year	1985	1987	1989	1994	1999	2004	2010	2015
Total number of fractures	677	773	996	1468	1697	2421	3218	3214
Man-to-woman ratio	1:2.7	1:2.4	1:2.8	1:2.9	1:3.2	1:3.6	1:3.9	1:4.0
Average age in male	67.5	70.4	71.4	74.4	75.5	77.8	78.9	81.4
Average age in female	76.2	76.9	77.7	80.9	80.5	83.3	83.7	84.9
Incidence (Fractures/100,000 per year)	132.1	138.8	179.7	230.0	236.2	314.4	385.0	370.9
Percentage of population > 65 years (%)	12.9	13.7	14.2	17.3	20.7	22.8	26.2	30.0

approach was advocated which would result in all patients receiving appropriate education and intervention. The notebooks were distributed to other hospitals in Niigata city that manage patients with hip fractures to increase adherence of treatment. Notably, implementation of FLS achieved higher adherence to drug therapy, lower subsequent fracture rate and mortality rate at 1 year. Our first report, which described the care of 167 hip fracture patients managed by the FLS, noted an adherence rate for osteoporosis treatment of 75.6%. At 1 year, the subsequent hip fracture was 1.8% and mortality rate was 7.3%. The Niigata FLS is a successful liaison service for secondary prevention of hip fracture.[5] Furthermore, Niigata Rehabilitation Hospital was recognised with the silver award from the IOF Capture the Fracture Best Practice Framework in 2016 representing the first such award in Japan. Finally, our most recent report published in 2017 observed that hip fracture incidence in Niigata prefecture had stopped increasing by 2015.[6] This report indicated that the hip fracture incidence rate was increasing until 2010, but incidence rates dropped form 385.0 in 2010 to 370.9 in 2015 per 100,000 of population per year (Table 6.1).

CONCLUSION

The implementation of systematic approaches to secondary fracture prevention in Japan has begun. However, health system barriers present major challenges to widespread establishment of FLS. Determined efforts are underway to identify the optimal model of care for secondary fracture prevention in Japan. More activity is needed to achieve a change in health-care policy and to drive greater awareness of the significance of fragility fractures among the Japanese population. We are seeking the optimal model of secondary fracture prevention approaches to improve patients care of fragility fracture. More efforts are required to increase public awareness of fragility fracture and to induce changes in health-care policy.

REFERENCES

1. Hagino H, et al. The risk of a second hip fracture in patients after their first hip fracture. *Calcif Tissue Int* 2012;**90**:14–21.
2. Miyakoshi N, et al. *J Jpn Orthop Assoc* 2012;**86**:913–20.
3. Kawashima T. Epidemiology of the femoral neck fracture in 1985, Niigata Prefecture, Japan. *J Bone Miner Metab* 1989;**7**:118–26.
4. Miyasaka D, et al. Incidence of hip fracture in Niigata, Japan in 2004 and 2010 and the long-term trends from 1985 to 2010. *J Bone Miner Metab* 2016;**34**:92–8.
5. Yamamoto N. Implementation of Fracture Liaison Service for a secondary prevention of fragility fracture. *Clin Calcium* 2017;**27**:1241–6.
6. Imai N, et al. A decrease in the number and incidence of osteoporotic hip fracture among elderly individuals in Niigata, Japan, from 2010 to 2015. *J Bone Miner Metab* 2018.
7. http://fragilityfracturenetwork.org/our-organisation/ffn-strategic-plan-2017-2021/.

Chapter 7

Secondary Fracture Prevention: Lebanon

Ghassan Maalouf, Maroun Rizkallah
Orthopedic Surgery Department, Bellevue University Medical Center, Saint-Joseph University, Beirut, Lebanon

INTRODUCTION

Fragility fractures are associated with increased morbidity and mortality in elderly patients, placing a large medical and economic burden on public healthcare systems.[1–3] Although age-specific rates of osteoporotic fractures are declining because of increased awareness, the absolute number of fractures continues to increase due to the ageing of the population.[4,5] The Middle East and especially Lebanon are currently witnessing an unprecedented increase in the number of older and elderly people.[6] Indeed, given the rapid decline in mortality and fertility as well as the emigration of younger people, projections predict that in 2025, about 10.2% of the Lebanese population will be 65 years or older.[7] Therefore, among other health issues, Lebanon will carry a large proportion of the osteoporosis burden in the coming decades.[8,9]

Despite the plethora of related publications and well-established guidelines for the care of patients with osteoporosis, the majority of patients sustaining a fragility fracture do not receive adequate osteoporosis treatment.[10] Worldwide, the rate of initiating osteoporosis treatment following a fragility fracture is low.[10–12] Preliminary yet unpublished surveys report a rate of 20% osteoporosis diagnosis and treatment initiation after a fragility fracture in some centres in Beirut. This rate is most probably lower in the remaining cities and other rural regions in Lebanon. Being aware of future changes in Lebanese demographics and the worldwide lack of secondary fracture prevention, many actions and plans were undertaken in Lebanon to increase awareness about fragility fracture care in a multitude of sectors.

EPIDEMIOLOGY OF FRAGILITY FRACTURES IN LEBANON

Sparse data from the medical literature show that the incidence of hip fractures in Lebanon in individuals over 50 years varies from 88 to 132 per 100,000 males

Secondary Fracture Prevention. https://doi.org/10.1016/B978-0-12-813136-7.00007-7

and from 160 to 188 per 100,000 females.[9,13] These studies showed an exponential increase in the risk of fracture with age and a trend towards a substantial increase in the incidence of hip fractures in Lebanon in upcoming decades.[9,13,14] Recent analyses put the prevalence of vertebral fracture in the Lebanese population at 26%.[15] Being aware of the burden of this disease, the Lebanese Society for Osteoporosis and Bone Metabolic Disorders issued the Lebanese guidelines for osteoporosis assessment and treatment, helping then in the primary prevention of fragility fractures.[16] The Lebanese Osteoporosis Prevention Society, an organisation actively involved in fracture prevention, strongly endorsed the Guidelines for Fragility Fractures in Lebanon, published in 2012 by Maalouf and colleagues. In their final recommendations, these guidelines clearly emphasise the necessity of implementing secondary fracture prevention strategies.[17] These strategies start with complete patient education and rehabilitation to reduce falls risk, assessment for bone fragility and evaluation of refracture risk and initiation of appropriate medical treatment for osteoporosis together with counselling for nutrition and dietary deficiencies.[17] All these measures can be completed through an organised system, the Fracture Liaison Service (FLS), aiming to 'make the first fracture the last'.[18,19] Secondary prevention in Lebanon is therefore working in three parallel sectors: The scientific contribution, the FLS initiative and the contribution by the health authorities.

SCIENTIFIC CONTRIBUTION

Secondary prevention starts by disseminating awareness, knowledge and prevention principles among the Lebanese scientific community, including orthopaedic surgeons, primary care physicians, rheumatologists, endocrinologists, allied health and nurses in hospitals and physical therapists. This was made possible by LOPS promoting regular integration of fragility fracture sessions at seminars and congresses for orthopaedic surgeons, gynaecologists, family and internal medicine physicians. These continuing medical education–accredited sessions are now fairly constant, and conferences about new osteoporosis treatments, Capture the Fracture, making the first fracture the last are regularly given. Sensitising the medical community to the concept of assessing and treating bone fragility after an index fracture was the first aim to be targeted through all conferences during 2016. Workshops are regularly organised by LOPS throughout Lebanon, encouraging physical therapists and nurses to disseminate knowledge about fall prevention exercises in patients at risk of osteoporotic fracture.

Evidence-based medicine being now the reference for medical practice, Lebanese practitioners published their work concerning osteoporosis and bone fragility. Available data on the epidemiology and health as well as economic burden of osteoporosis in Lebanon highlight the importance of prevention, whether primary or secondary.[9,13–16] Maalouf et al. and Rizkallah et al. emphasised the importance of secondary prevention and designed a road map to its application throughout Lebanon.[17,20]

Furthermore, partnership and regular participation in international meetings of societies and organisations such as International Osteoporosis Foundation (IOF) and the Fragility Fracture Network are regularly accomplished in collaboration with the Lebanese Order of Physicians to keep our medical committees up to date with the newest international guidelines and strategies. This is an essential step to benefit from the expertise of these specialised organisations, especially during the bone and joint decade.

FRACTURE LIAISON SERVICE

FLS is a special program designed to identify, investigate and initiate appropriate treatment for patients presenting with a fragility fracture, identified as having compromised bone quality, who therefore are at high risk for secondary fractures.[21–23] Overwhelming evidence supports the role of FLS in increasing rates of dual-energy X-ray absorptiometry (DEXA) screening and initiation of osteoporosis treatment after an index fracture.[24] Some reports also confirmed a reduction in the rate of refractures following the implementation of an FLS, demonstrating the clinical- and patient-oriented benefit of this system.[25–27] A thorough medical literature review produced no reports on FLS in the Middle East, while the IOF is trying to convince Arab health-care systems to introduce the FLS model in their hospitals. However, in a Lebanese initiative, a type A FLS was created at Bellevue Medical Center, Beirut, in July 2013. After overcoming many obstacles, this FLS is now fully functioning and enjoys the collaboration of all medical and nursing staff in the hospital. Patients are satisfied with the service, and good adherence to recommendations has been observed. The IOF contacted Bellevue hospital and is now collaborating with the medical administration to share this practice with other local and regional hospitals to convince them of the importance of such an experience. To create solid evidence, a scientific study is being conducted evaluating the efficacy of this system in Lebanon, with preliminary results showing significant improvement in osteoporosis screening and initiation of treatment, as well as refracture incidence rates. These results, when published, will constitute a solid ground for the application of the Beirut FLS model throughout Lebanese hospitals. The FLS model being a pillar in the secondary fracture prevention, its diffusion will take prevention of fragility fractures in Lebanon and the surrounding countries to the next level.

CONTRIBUTIONS BY THE HEALTH AUTHORITIES

Influence and input from LOPS and the Lebanese Order of Physicians led to many reparative measures in the prevailing Lebanese Health System. The Lebanese Ministry of Public Health (MoPH) is now conscious of the heavy economic and health burden of fragility fractures and is doing the needful to reduce it. Since January 2016, the Lebanese Government agreed on a plan providing full financial hospital care coverage for patients aged 65 years or older who present with a fragility fracture requiring hospitalisation and surgical care. The MoPH also launched the National Campaign for Osteoporosis Awareness in 2015; this included extensive press, radio and TV conferences addressing the public, as well as posters along public roads encouraging people with clinical risk factors for osteoporosis to ask their doctor for a bone health assessment. The campaign also included a month of waved fees for osteodensitometry (DEXA bone scan).

Being conscious of the economic benefit of treating osteoporosis after an index fracture rather than treating its complications (the refracture), the MoPH offers free first- and second-line osteoporosis treatments as indicated, pushing patients to be evaluated after the index fracture and therefore to be treated, minimising the risk of a second fracture after the first fracture. Knowing that indexing and archiving is essential for accurate statistics and that statistical analysis is paramount for showing gaps in the health-care system, the MoPH is now also working on issuing the full hip fracture registry in Lebanon. The MoPH is holding regular meeting with LOPS delegated by the Lebanese Order of Physicians to evaluate the actual situation and to discuss plans for improvement, including sponsoring a National scientific meeting that will be held in 2018 to diffuse the FLS experience.

CONCLUSION

The recommendations by Maalouf et al.[17] and Rizkallah et al.[20] were a big step towards a better understanding and hence improved secondary prevention of fragility fracture in Lebanon. Much still needs to be done. For the upcoming years, the plan is to appear in multispecialists society meetings to talk about fragility fracture prevention, to increase the number of studies providing a high level of evidence and to actively participate in international organisations and congresses. Our aim for 2018 is to expand the successful Beirut FLS model to all Lebanese territories, making use of our solid and undebatable evidence that this model leads to a reduction in the risk of refracture. LOPS will keep its regular meetings with the Lebanese health-care authorities to continuously assess the prevention status and intervene on any lack to the standardised worldwide guidelines. Computerised registries will be available to statisticians for regular checking and analysis so we can find the gaps and bridge them.

REFERENCES

1. Rosengren BE, Karlsson M, Petersson I, Englund M. The 21st-century landscape of adult fractures: cohort study of a complete adult regional population. *J Bone Miner Res* March 2015;**30**(3):535–42. [Internet] Available from: http://www.ncbi.nlm.nih.gov/pubmed/25280349.
2. Ballane G, Cauley JA, Luckey MM, Fuleihan GE-H. Secular trends in hip fractures worldwide: opposing trends East versus West. *J Bone Miner Res* August 2014;**29**(8):1745–55. [Internet] Available from: http://www.ncbi.nlm.nih.gov/pubmed/24644018.

3. Cauley JA, Chalhoub D, Kassem AM, Fuleihan GE-H. Geographic and ethnic disparities in osteoporotic fractures. *Nat Rev Endocrinol* June 2014;**10**(6):338–51. [Internet] Available from: http://www.ncbi.nlm.nih.gov/pubmed/24751883.
4. Willson T, Nelson SD, Newbold J, Nelson RE, LaFleur J. The clinical epidemiology of male osteoporosis: a review of the recent literature. *Clin Epidemiol* 2015;**7**:65–76. [Internet] Available from: http://www.ncbi.nlm.nih.gov/pubmed/25657593.
5. Curran D, Maravic M, Kiefer P, Tochon V, Fardellone P. Epidemiology of osteoporosis-related fractures in France: a literature review. *Joint Bone Spine* December 2010;**77**(6):546–51. [Internet] Available from: http://www.ncbi.nlm.nih.gov/pubmed/20378383.
6. Hajjar RR, Atli T, Al-Mandhari Z, Oudrhiri M, Balducci L, Silbermann M. Prevalence of aging population in the Middle East and its implications on cancer incidence and care. *Ann Oncol* October 2013;**24**(Suppl. 7):vii11–24. [Internet] Available from: http://www.ncbi.nlm.nih.gov/pubmed/24001758.
7. Sibai AM, Sen K, Baydoun M, Saxena P. Population ageing in Lebanon: current status, future prospects and implications for policy. *Bull World Health Organ* March 2004;**82**(3):219–25. [Internet] Available from: http://www.ncbi.nlm.nih.gov/pubmed/15112011.
8. Mehio Sibai A, Rizk A, Kronfol NM. Aging in Lebanon: perils and prospects. *J Med Liban* 2015;**63**(1):2–7. [Internet] Available from: http://www.ncbi.nlm.nih.gov/pubmed/25906507.
9. Maalouf G, Bachour F, Hlais S, Maalouf NM, Yazbeck P, Yaghi Y, et al. Epidemiology of hip fractures in Lebanon: a nationwide survey. *Orthop Traumatol Surg Res* October 2013;**99**(6):675–80. [Internet] Available from: http://www.ncbi.nlm.nih.gov/pubmed/24007698.
10. Byszewski A, Lemay G, Molnar F, Azad N, McMartin SE. Closing the osteoporosis care gap in hip fracture patients: an opportunity to decrease recurrent fractures and hospital admissions. *J Osteoporos* 2011;**2011**:404969. [Internet] Available from: http://www.ncbi.nlm.nih.gov/pubmed/21977330.
11. Ekman EF. The role of the orthopaedic surgeon in minimizing mortality and morbidity associated with fragility fractures. *J Am Acad Orthop Surg* May 2010;**18**(5):278–85. [Internet] Available from: http://www.ncbi.nlm.nih.gov/pubmed/20435878.
12. Leslie WD, Giangregorio LM, Yogendran M, Azimaee M, Morin S, Metge C, et al. A population-based analysis of the post-fracture care gap 1996-2008: the situation is not improving. *Osteoporos Int* May 2012;**23**(5):1623–9. [Internet] Available from: http://www.ncbi.nlm.nih.gov/pubmed/21476038.
13. Sibai AM, Nasser W, Ammar W, Khalife MJ, Harb H, Fuleihan GE-H. Hip fracture incidence in Lebanon: a national registry-based study with reference to standardized rates worldwide. *Osteoporos Int* September 2011;**22**(9):2499–506. [Internet] Available from: http://www.ncbi.nlm.nih.gov/pubmed/21069293.
14. Baddoura R, Hoteit M, El-Hajj Fuleihan G. Osteoporotic fractures, DXA, and fracture risk assessment: meeting future challenges in the Eastern Mediterranean Region. *J Clin Densitom* 2011;**14**(4):384–94. [Internet] Available from: http://www.ncbi.nlm.nih.gov/pubmed/21839659.
15. Ballane G, Cauley JA, Luckey MM, El-Hajj Fuleihan G. Worldwide prevalence and incidence of osteoporotic vertebral fractures. *Osteoporos Int* May 2017;**28**(5):1531–42. [Internet] Available from: http://www.ncbi.nlm.nih.gov/pubmed/28168409.
16. El-Hajj Fuleihan G, Baddoura R, Awada H, Arabi A, Okais J. First update of the Lebanese guidelines for osteoporosis assessment and treatment. *J Clin Densitom* 2008;**11**(3):383–96. [Internet] Available from: http://www.ncbi.nlm.nih.gov/pubmed/18448373.
17. Maalouf G, Bachour F, Issa M, Yazbeck P, Maalouf N, Daher C, et al. Guidelines for fragility fractures in Lebanon. *J Med Liban* 2012;**60**(3):153–8. [Internet] Available from: http://www.ncbi.nlm.nih.gov/pubmed/23198456.

18. Akesson K, Marsh D, Mitchell PJ, McLellan AR, Stenmark J, Pierroz DD, et al. Capture the Fracture: a best practice framework and global campaign to break the fragility fracture cycle. *Osteoporos Int* August 2013;**24**(8):2135–52. [Internet] Available from: http://www.ncbi.nlm.nih.gov/pubmed/23589162.
19. Eisman JA, Bogoch ER, Dell R, Harrington JT, McKinney RE, McLellan A, et al. Making the first fracture the last fracture: ASBMR task force report on secondary fracture prevention. *J Bone Miner Res* October 2012;**27**(10):2039–46. [Internet] Available from: http://www.ncbi.nlm.nih.gov/pubmed/22836222.
20. Rizkallah M, Sebaaly A. Commentary: where are we from the implementation of fragility fracture guidelines in Lebanon?. *J Yoga Phys Ther* 2016;**6**(3). [Internet] Available from: http://www.omicsonline.org/open-access/commentary-where-are-we-from-the-implementation-of-fragility-fractureguidelines-in-lebanon-2157-7595-1000244.php?aid=75049.
21. Huntjens KMB, van Geel TACM, van den Bergh JPW, van Helden S, Willems P, Winkens B, et al. Fracture liaison service: impact on subsequent nonvertebral fracture incidence and mortality. *J Bone Joint Surg Am* February 19, 2014;**96**(4):e29. [Internet] Available from: http://www.ncbi.nlm.nih.gov/pubmed/24553898.
22. Miller AN, Lake AF, Emory CL. Establishing a fracture liaison service: an orthopaedic approach. *J Bone Joint Surg Am* April 15, 2015;**97**(8):675–81. [Internet] Available from: http://www.ncbi.nlm.nih.gov/pubmed/25878314.
23. Marsh D, Akesson K, Beaton DE, Bogoch ER, Boonen S, Brandi M-L, et al. Coordinator-based systems for secondary prevention in fragility fracture patients. *Osteoporos Int* July 2011;**22**(7):2051–65. [Internet] Available from: http://www.ncbi.nlm.nih.gov/pubmed/21607807.
24. Walters S, Khan T, Ong T, Sahota O. Fracture liaison services: improving outcomes for patients with osteoporosis. *Clin Interv Aging* 2017;**12**:117–27. [Internet] Available from: http://www.ncbi.nlm.nih.gov/pubmed/28138228.
25. Nakayama A, Major G, Holliday E, Attia J, Bogduk N. Evidence of effectiveness of a fracture liaison service to reduce the re-fracture rate. *Osteoporos Int* March 2016;**27**(3):873–9. [Internet] Available from: http://www.ncbi.nlm.nih.gov/pubmed/26650377.
26. Lih A, Nandapalan H, Kim M, Yap C, Lee P, Ganda K, et al. Targeted intervention reduces refracture rates in patients with incident non-vertebral osteoporotic fractures: a 4-year prospective controlled study. *Osteoporos Int* March 2011;**22**(3):849–58. [Internet] Available from: http://www.ncbi.nlm.nih.gov/pubmed/21107534.
27. Van der Kallen J, Giles M, Cooper K, Gill K, Parker V, Tembo A, et al. A fracture prevention service reduces further fractures two years after incident minimal trauma fracture. *Int J Rheum Dis* February 2014;**17**(2):195–203. [Internet] Available from: http://www.ncbi.nlm.nih.gov/pubmed/24576275.

Chapter 8

Fracture Liaison Services in South East Asia: Notes From a Large Public Hospital in Singapore

Manju Chandran
Osteoporosis and Bone Metabolism Unit, Department of Endocrinology, Singapore General Hospital, Singapore

INTRODUCTION

The knowledge that fractures beget fractures has been capitalised upon in the concept of fracture liaison services (FLSs).[1–5] Organisations such as the International Osteoporosis Foundation and the American Society for Bone and Mineral Research have formed task forces and launched campaigns that provide best practice frameworks, resources and toolkits for the establishment of effective FLS.[6,7] The recent implementation by the IOF Capture the Fracture Best Practice Framework, of a method of global benchmarking of FLS around the world has, however, confirmed the heterogeneity of such services and highlights the fact that translating principles of exemplary medical practice into daily patient care is not easy.[8] Socioeconomic and geopolitical barriers to successful implementation of such services may exist in various parts of the globe and this has to be considered when implementing a FLS.

THE SINGAPORE STORY

An understanding of the health-care environment and methods of provision of medical care in Singapore is essential to appreciate the background under which OPTIMAL (Osteoporosis Patient Targeted and Integrated Management for Active Living), the FLS funded by the Ministry of Health (MOH), came into being in 2008.

Singapore, a small island country in South East Asia, has a population of 5.6 million, made up of people of multiple races and cultural backgrounds with the majority being Chinese (74.2%) and the rest being of Malay (13.2%),

Secondary Fracture Prevention. https://doi.org/10.1016/B978-0-12-813136-7.00008-9

Indian (9.2%) and other mixed ethnicities (http://www.singstat.gov.sg/statistics). Health care in Singapore is under the responsibility of the Singapore Government's MOH. Subsidies on health care are provided, financed through taxation and through a nationalised health insurance plan known as Medisave. Medisave makes use of compulsory savings deducted from payrolls of citizens and residents. Through Medisave, each person accumulates funds that are individually tracked, and such funds can be pooled within and across an entire extended family. A key principle of the national health scheme is that, even in the public health-care system, though it may be heavily subsidised, no medical service is provided completely free of charge. Out-of-pocket charges vary considerably for each service and level of subsidy. Complementary schemes such as Medishield, Medifund and Eldershield also exist.

70%–80% of Singaporeans obtain their medical care within the public health system.[9] The public acute general hospitals are referred to as 'restructured hospitals', and they operate as government-owned corporations that receive a government subsidy for the provision of subsidised medical services to their patients. Each public hospital is a legally autonomous entity registered as a private firm. Though owned by the government and therefore ultimately accountable to it, unlike other governmental organisations, each public hospital has operational autonomy in all areas, including recruitment, remuneration, purchase and pricing of services. Primary health care is provided by a mix of 18 government outpatient primary care centres (polyclinics) and 2000 private medical practitioners' clinics. Polyclinics are heavily subsidised and have onsite investigative facilities and pharmacy services. They provide outpatient care for both acute and chronic medical conditions.

A Chronic Disease Management Program (CDMP) exists under which 10 diseases are included and patients can utilise Medisave up to a fixed amount of SGD$ 400/year to reduce out-of-pocket cash payments for outpatient bills associated with these diseases. Osteoporosis was included under the CDMP in July 2015.

Singapore has a rapidly aging population with the number of people aged 65 years and older expected to triple from the current 350,000 to 960,000 by the year 2030. The age adjusted rates of osteoporosis among women over the age of 50 years in Singapore are currently among the highest in Asia.[10] The direct costs imposed by osteoporotic fractures in Singapore are very high, closely paralleling that of Western countries.[11] Mortality rates following hip fractures[12] in Singapore as well as rehabilitation potential[13] in patients who survive are similar to that of Western industrialised nations. Only about 1 in 3 people who survive a hip fracture return to their previous level of independence, 50% require long-term help with routine activities and cannot walk unaided, whereas 25% require full-time nursing home care.[13] The main social and economic burden is borne by the families of those affected.[14]

OPTIMAL is a secondary fracture prevention programme funded by the Singapore MOH and it was set up in the public acute general hospitals and polyclinics of Singapore in 2008. At that time, the public health-care system comprised of only two large 'clusters' and both were involved in OPTIMAL. (These 'clusters' were subsequently organised into six regional health systems and soon will be reorganised into three integrated 'clusters'). The aim of OPTIMAL was to target every Singaporean patient above 50 years of age who had suffered a previous fragility fracture. These patients would be evaluated for osteoporosis and falls risk and provided with an integrated CDMP for secondary fracture prevention. Provisions were offered through OPTIMAL for case finding, performing and assessing diagnostic evaluations (including axial DXA), providing recommendations for falls prevention, prescribing exercise and making specific treatment recommendations regarding antiosteoporosis pharmacotherapy.

Findings from an earlier pilot project, the HSDP Osteoporosis Management Program – conducted in one of the two health-care clusters, generated the evidence base required to convince policymakers in the MOH of the importance of a fracture prevention programme and to provide funding for the implementation of OPTIMAL. The HSDP project recruited a total of 1056 patients across three hospitals and nine polyclinics over a 2-year period and demonstrated that a significant care gap existed in osteoporosis with only 16% of patients found to have been started on appropriate treatment for their osteoporosis within 2 years of their hip fracture. Prior to implementing OPTIMAL, designated clinician champions from the various hospitals met up for several rounds of discussions, and certain general standards of care in accordance with evidence-based national guidelines were adopted. The targeted key performance indicators (KPIs) set to be achieved at a 2-year follow-up mark were as follows:

1. A 40% reduction in refracture incidence,
2. An 85% increase in the proportion of patients who have evaluation for osteoporosis and fracture risk after sustaining a fracture and receive appropriate treatment when indicated and
3. A 70% adherence rate to medications.

In 2008, generic bisphosphonates were not yet available in Singapore. An agreement with the MOH was made for subsidising the following medications – oral Alendronate (Fosamax 70 mg/week), oral Risedronate (Actonel 35 mg/week), intravenous Zoledronic Acid (Aclasta 5 mg/year) and oral daily Strontium Ranelate (Protos 2 mg/day) for the first 2 years of the programme. After the first 2 years, pharmaceutical companies worked with individual hospitals to provide discounts for patients recruited through the OPTIMAL programme.

Recognising the vital importance of a dedicated case manager at the core of such a FLS, clear job descriptions for potential hires were drawn up and used in the job recruitment process for these individuals at all the hospitals. The projected workload of the full-time employed case manager was calculated such that he or she would be expected to see, screen, evaluate and provide osteoporosis and falls risk assessment to eight patients per half working day. Given that the modus operandi of the different hospitals varied, individual clinician champions were given the 'carte blanche' to implement the programme in their hospitals in a way that would work best at their respective institutions.

With 1600 beds, Singapore General Hospital is the largest tertiary teaching and public hospital in Singapore. It is the flagship hospital of the SingHealth cluster of medical institutions. The hospital on average sees 600 patients with hip fractures admitted through its Emergency department per year. A dedicated Osteoporosis and Bone Metabolism Unit under the department of Endocrinology with medical, nursing, dietetics and physiotherapy services was set up in 2007 at the LIFE Centre (Lifestyle and Fitness Enhancement Centre), though patients with osteoporosis are seen and are offered care also in multiple other departments including the Departments of Orthopedics, Rheumatology, Internal Medicine, Obstetrics and Gynecology and Physical Medicine and Rehabilitation. The hospital has an electronic patient records management system (Sunrise Clinical Manager) and a pharmacy patient purchase records system (MAXCARE). The Accident and Emergency Department makes use of an electronic medical record known as EMERGE for managing data entry, storage and retrieval of patient-specific information, and it uses the current ICD codes for diagnostic coding. Inpatient discharges are documented through the HIDS (Hospital Inpatient Discharge System). The hospital also makes use of the Singapore national electronic health record system (Electronic Medical Record Exchange (EMRX)) that allows all public hospitals and polyclinics in Singapore to share patient records online.

At Singapore General Hospital, OPTIMAL was set up as a multidisciplinary effort between the Osteoporosis and Bone Metabolism unit of the Department of Endocrinology and the multiple other departments that see patients with osteoporosis and osteoporotic fractures. At the time of commencement of the programme, preliminary steps by the Osteoporosis and Bone Metabolism Unit to implement an 'Osteoporosis Prevention and Treatment Initiative' at the hospital had already been taken and the existing hip fracture pathway of the department of orthopedics had been redesigned and updated in consultation with the Quality Improvement Committee. This made it easier for the programme to hit the ground running. We set the programme up in such a way that a case manager specific to the programme would be informed each time that a physician from any department within the hospital identified patients with a low trauma fracture who were 50 years of age or older. The case manager is a specialist nurse and he/she assesses both in- and outpatients referred to the service. Fracture case records from the Accident and Emergency department are also screened on a biweekly basis to identify patients with low trauma fractures that are seen in the department and these patients are then contacted to see if they would be willing to be recruited into the programme. The case manager thus serves not only as a 'referral person' but also as a 'case finder'. Recommendations are made to the referring physician to order in all their patients who consent to be in the programme, a basic laboratory work up to rule out common secondary contributors to osteoporosis that includes a full blood count, bone profile including calcium, phosphate, alkaline phosphatase and vitamin D levels, thyroid function tests (TSH and free thyroxine) and serum creatinine. DXA of the hip and spine is recommended to be done in all patients at baseline and at the end of 2 years to facilitate monitoring of changes in bone mineral density change. At our hospital, the ultimate choice of which therapy to start the patient on is left to the discretion of the referring physician. Falls risk assessment is performed on all patients recruited into the programme. Patients who are not wheel chair bound are encouraged to be enrolled into one of three exercise programmes – the OTAGO, a group strength and balance retraining programme designed to prevent falls in older people living in the community or individual physiotherapy exercise sessions or community-based exercise (e.g., Tai Chi). If a decision is made to discharge the patient from the Specialist Outpatient Clinics at the hospital to the polyclinic, the OPTIMAL case manager at the hospital liaises with his/her counterpart at the particular polyclinic that is most easily accessible to the patient and transfers follow-up care of the patient to that clinic. This is through telephonic transmission of information regarding the patient to the polyclinic OPTIMAL case manager and by faxing over a discharge summary of the patient to the polyclinic.

OPTIMAL maintains a centralised computerised data base (CCRD) for the entry of all baseline data including demographic details, past medical and surgical history including that of fractures and falls, risk factors for osteoporosis, current medication use, dietary calcium intake, DXA results and interpretation, lifestyle and treatment recommendations and arrangements for follow up. Detailed operational characteristics of the OPTIMAL programme at Singapore General Hospital have already been described elsewhere.[3] The criteria for inclusion of patients into OPTIMAL are shown in Table 8.1. The components and the workflow of OPTIMAL are shown in Figs. 8.1 and 8.2, respectively.

TABLE 8.1 Inclusion Criteria for Osteoporosis Patient Targeted and Integrated Management for Active Living
Male or Female >50-years old
Patient with previous vertebral or nonvertebral fragility fracture (except skull/facial fracture and fracture below ankle or distal to wrist)
Agree to participate in the programme and agree to follow up for at least 2 years

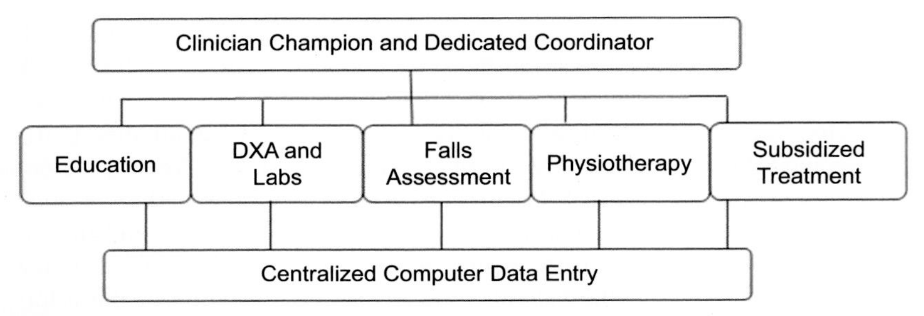

FIGURE 8.1 Components of Osteoporosis Patient Targeted and Integrated Management for Active Living.

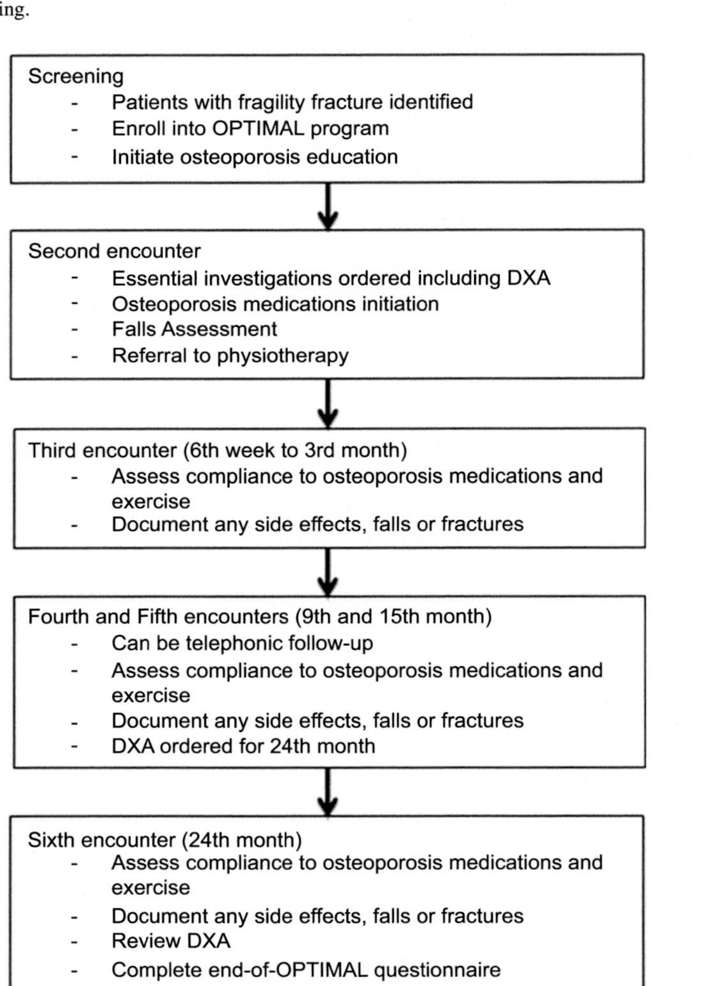

FIGURE 8.2 Workflow of Osteoporosis Patient Targeted and Integrated Management for Active Living (OPTIMAL).

STRENGTHS OF OSTEOPOROSIS PATIENT TARGETED AND INTEGRATED MANAGEMENT FOR ACTIVE LIVING

OPTIMAL is a highly facilitated and multipronged programme. Both the care coordinator and the clinician champion play vital roles in it. While the role of the care coordinator is primarily in ensuring a smooth daily 'on the ground' running of the programme, clinician champions who are committed, enthusiastic and persistent, will lead service development, are able to work with senior management and governmental bodies to secure funding, are able to convince busy clinicians in multiple departments to 'buy in' to a programme that may involve extra short-term work and are dedicated enough to devote time and effort to maintain the service are an absolute necessity in any FLS. Delivery processes that include serial clinical interventions, namely case finding, bone density measurement, evaluation of secondary causes, initiation of pharmacological and nonpharmacological treatments and methods to ensure treatment compliance had to be outlined and put into place at the beginning itself in our programme and this would not have been possible without the commitment of a persistent and passionate clinician champion. The clinician champion also should be able to keep up with the evolving and constantly changing health-care landscape and to maintain constant ties with management within the hospital as well as policymakers and governmental bodies and to appraise them of issues faced while running the programme. Having had the programme endorsed at the national level before it was instituted at our hospital made it easier for the OPTIMAL team to obtain the acknowledgement and help of senior leadership within our hospital. Leveraging on the electronic prescription, medication refill and medical records that were available in our health-care system as well as the CCRD that is part of the OPTIMAL framework to capture all pertinent data related to our patients has helped us capture all pertinent data in its entirety. This has helped us when auditing data.

In 2014, we conducted an internal audit of patients who had completed at least 2 years of follow-up after being recruited into OPTIMAL in our hospital. The findings of our audit showed good osteoporosis assessment and treatment rates that were in line with the KPIs laid down at the start of the programme and described earlier.[15] The findings from our audit also indicated that a high proportion of patients maintained good compliance to treatment over the 2 year follow-up period. Mean Medication Possession Ratio (MPR) $\pm$ SD at 2 years was 72.8 $\pm$ 34.5%. A Kaplan–Meier survival analysis showed that the proportion of patients with a MPR of $\geq$80% at 12, 18 and 24 months of follow-up was 83%, 75% and 50%, respectively.[3] The reason behind the good compliance to therapy is likely largely due to the tracking of patients' progress at regular intervals and close follow-up by the case manager. This observation has been validated in other studies also.[16]

WHERE HAVE WE 'DROPPED THE BALL'?

Although the quality benchmarking of the Best Practice Framework of the International Osteoporosis Foundation's Capture the Fracture programme honoured the OPTIMAL programme at Singapore General Hospital with a Gold Star in 2015,[8] we are fully cognizant of the fact that there are still several drawbacks to our programme and that there are multiple challenges to overcome. The process of implementing OPTIMAL and running it over the last 9 years at the largest hospital in Singapore has helped us to identify these problems and limitations and have served as valuable learning points which we hope to leverage on to improve the programme. Implementing a FLS is only half the battle. It is critical whilst running such services to look back and evaluate where, as providers of health care, we could have 'dropped the ball' and why patients could have 'fallen off the care wagon'.

It is apparent that there are both patient-level and system-level challenges and barriers to successful maintenance of a FLS. In 2016, we published one of the first studies that explored patient-related factors that led them to drop off our secondary fracture prevention programme.[17] We found that of 938 patients followed up over 2 years, 237 defaulted at various time points. A significant percentage of patients who dropped out of the programme opined that it was because the follow-up visits were too time consuming. An interesting observation from our patient population that comprises of multiple ethnicities was that non-Chinese patients compared to Chinese (aHR = 1.98, 1.33–2.94), patients with primary school education and below compared to those with secondary school and above education (aHR = 1.65, 1.11–2.45), and those with nonvertebral and/or multiple fractures compared to those with spine fractures (aHR = 1.38, 1.06–1.81), were more likely to be nonadherent. A detailed discussion and analysis of these patient-level challenges can be obtained from our published article.[17] System-level failures inability to attain a 100% 'capture rate' of all patients with fragility fractures presenting to our large hospital. This problem can partly be attributed to the lack of adequate number of personnel to serve as case managers and this may be partially due to manpower funding issues as well as the lack of understanding still persistent amongst hiring personnel and junior-level staff of the importance and long-term value of such a secondary fracture prevention programme and their critical role in it. It is also a well-recognised fact that some osteoporotic fractures especially vertebral fractures go often undiagnosed.[18] Recognising this deficit, we are currently working to put in place modalities such as Vertebral Fracture Assessment concurrently with the performance of DXA scans to identify the occult vertebral fracture population.[19] Right-siting,[20] a term unique to Singapore health-care parlance, i.e., discharging patients to polyclinics and GP practices after initial assessment and therapy initiation also remains inadequate in our programme as does compliance to the exercise and fall prevention services offered through the programme.

WHERE DO WE GO FROM HERE?

The challenges identified during Phase 1 of OPTIMAL has led to plans and proposals for a better and improved Phase 2 at a nationwide level. Amongst these are included plans to integrate OPTIMAL with Orthogeriatric and/or Hip Fracture Care[21] pathways that already exist in almost all of the public acute care hospitals in Singapore and are linked to community step down care facilities. We also plan to integrate not only the currently included government polyclinic physicians but also private GPs into the programme. We are planning to provide not only more seamless transition pathways with the care coordinators liaising with the GPs after initial diagnosis and treatment of the patient at the hospital but also to involve them at the outset itself by training them in Osteoporosis care and its diagnostic and medical management. Plans to develop community-based exercise and fall prevention programmes that will overcome the hurdle of poor compliance to such programmes that currently run out of only the hospitals are also in the offing. Rather than continue to have OPTIMAL remain as a stand-alone project that depends on continued outside funding whether it be from governmental or other bodies, proposals to integrate OPTIMAL into the standard workflow of all the hospitals in Singapore so that it forms part of their routine practice and enable 'block funding', i.e., to fund the FLS at each hospital as part of the operating budget of the hospital are also being actively explored.

CONCLUSION

FLSs are the new face of osteoporotic fracture care, providing an invaluable service to patients. The initial positive results from the first phase of OPTIMAL in Singapore and the lessons learned during the last several challenging but satisfying years has set the stage for us to implement the second phase of OPTIMAL here. It is also our goal to demonstrate the cost effectiveness of the programme and to make sure that no patient with a fragility fracture is ever neglected and that the first fracture if at all it happens in any patient is really the last.

ACKNOWLEDGEMENTS

The author would like to acknowledge the OPTIMAL Care Coordinators at Singapore General Hospital who have worked with her over the last 9 years for their dedication and commitment to secondary fracture prevention as well as the continuous support and counsel of Dr. Lau Tang Ching-Project Director of OPTIMAL, Singapore.

REFERENCES

1. McLellan AR, Gallacher SJ, Fraser M, McQuillian C. The fracture liaison service: success of a program for the evaluation and management of patients with osteoporotic fracture. *Osteoporos Int* 2003;**14**(12):1028–34.
2. Lih A, Nandapalan H, Kim M, Yap C, Lee P, Ganda K, et al. Targeted intervention reduces refracture rates in patients with incident non-vertebral osteoporotic fractures: a 4-year prospective controlled study. *Osteoporos Int* 2011;**22**(3):849–58.
3. Chandran M, Tan MZW, Cheen M, Tan SB, Leong M, Lau TC, et al. Secondary prevention of osteoporotic fractures-an "OPTIMAL" model of care from Singapore. *Osteoporos Int* 2013;**24**(11):2809–17.

4. Dell R. Fracture prevention in Kaiser Permanente Southern California. *Osteoporos Int* 2011;**22**(Suppl. 3):457–60.
5. Mitchell P, Åkesson K, Chandran M, Cooper C, Ganda K, Schneider M, et al. Implementation of Models of Care for secondary osteoporotic fracture prevention and orthogeriatric Models of Care for osteoporotic hip fracture. *Best Pract Res Clin Rheumatol* 2016;**30**(3):536–58.
6. Akesson K, Marsh D, Mitchell PJ, McLellan AR, Stenmark J, Pierroz DD, et al. Capture the fracture: a best practice framework and global campaign to break the fragility fracture cycle. *Osteoporos Int* 2013;**24**(8):2135–52.
7. Eisman JA, Bogoch ER, Dell R, Harrington JT, McKinney RE, McLellan AL, et al. Making the first fracture the last fracture ASBMR task force report on secondary fracture prevention. *J Bone Miner Res* 2012;**27**(10):2039–46.
8. Javaid MK, Kyer C, Mitchell PJ, Chana J, Moss C, Edwards MH, et al. Effective secondary fracture prevention: implementation of a global benchmarking of clinical quality using the IOF Capture the Fracture® Best Practice Framework tool. *Osteoporos Int* 2015;**26**(11):2573–8.
9. Lim MK. Transforming Singapore health care: public-private partnership. *Ann Acad Med Singap* 2005;**34**(7):461–7.
10. Lau EM, Lee JK, Suriwongpaisal P, Saw SM, Das De S, Khir A, et al. The incidence of hip fracture in four Asian countries: the Asian osteoporosis study (AOS). *Osteoporos Int* 2001;**12**(3):239–43.
11. Ng CS, Lau TC, Ko Y. Cost of osteoporotic fractures in Singapore. *Value Health Reg Issues* 2017;**12**:27–35.
12. Tay E. Hip fractures in the elderly: operative versus nonoperative management. *Singap Med J* 2016;**57**(4):178–81.
13. Pasco JA, Sanders KM, Hoekstra FM, Henry MJ, Nicholson GC, Kotowicz MA, et al. The human cost of fracture. *Osteoporos Int* 2005;**16**(12):2046–52.
14. Siddiqui MQ, Sim L, Koh J, Fook-Chong S, Tan C, Howe TS, et al. Stress levels amongst caregivers of patients with osteoporotic hip fractures – a prospective cohort study. *Ann Acad Med Singap* 2010;**39**(1):38–42.
15. Cheen MHH, Kong MC, Zhang RF, Tee FMH, Chandran M. Adherence to osteoporosis medications amongst Singaporean patients. *Osteoporos Int* 2012;**23**(3):1053–60.
16. Bogoch ER, Elliot-Gibson V, Beaton D, Sale J, Josse RG. Fracture prevention in the orthopaedic environment: outcomes of a coordinator-based fracture liaison service. *J Bone Joint Surg Am* 2017;**99**(10):820–31.
17. Chandran M, Cheen M, Ying H, Lau TC, Tan M. Dropping the ball and falling off the care wagon. Factors correlating with nonadherence to secondary fracture prevention programs. *J Clin Densitom* 2016;**19**(1):117–24.
18. Delmas PD, van de Langerijt L, Watts NB, Eastell R, Genant H, Grauer A, et al. Underdiagnosis of vertebral fractures is a worldwide problem: the IMPACT study. *J Bone Miner Res* 2005;**20**(4):557–63.
19. Jager PL, Slart RHJA, Webber CL, Adachi JD, Papaioannou AL, Gulenchyn KY, et al. Combined vertebral fracture assessment and bone mineral density measurement: a patient-friendly new tool with an important impact on the Canadian Risk Fracture Classification. *Can Assoc Radiol J* 2010;**61**(4):194–200.
20. Wee SL, Tan CGP, Ng HSH, Su S, Tai VUM, Flores JVPG, et al. Diabetes outcomes in specialist and general practitioner settings in Singapore: challenges of right-siting. *Ann Acad Med Singap* 2008:37.
21. Tan LTJ, Wong SJ, Kwek EBK. Inpatient cost for hip fracture patients managed with an orthogeriatric care model in Singapore. *Singap Med J* 2017;**58**(3):139–44.

Chapter 9

Fracture Liaison Services in Taiwan: Developments and Perspectives

Lo-Yu Chang[1], **Ding-Cheng (Derrick) Chan**[2,3,4]
[1]School of Medicine, National Taiwan University, Taipei, Taiwan; [2]Department of Geriatrics and Gerontology, National Taiwan University Hospital, Taipei, Taiwan; [3]Department of Internal Medicine, National Taiwan University Hospital, Taipei, Taiwan; [4]Superintendent Office, Chu-Tung Branch, National Taiwan University Hospital, Hsinchu County, Taiwan

INTRODUCTION

Previous studies have shown that, worldwide, less than one-third of patients with fragility fractures receive adequate osteoporosis evaluation and treatment.[1] This widespread failure in medical care provision triggered the development and implementation of secondary fracture prevention programs (aka Fracture Liaison Services (FLSs)) in a number of countries, namely the United Kingdom (Glasgow), the United States (California), New Zealand and Australia (Sydney). The success of secondary fracture prevention programs in these countries highlights the importance of secondary fracture prevention program in osteoporosis management.[2–4] In 2012, the International Osteoporosis Foundation (IOF) added the 'Capture the Fracture' campaign to help medical institutions everywhere establish FLSs. The 'Best Practice Framework' (BPF) specified 13 standards to provide guidance for the relevant institutions. As a result, fracture prevention services have been able to increase medication adherence,[5–7] reduce the risk of refracture[7,8] and decrease mortality[7,8] with reasonable cost-effectiveness.[8–11]

The annual incidence of hip fracture in Taiwan is among the highest in the world: The 2010 United Nation world population adjusted incidence was 392/100,000 for women and 196/100,000 for men, respectively.[12,13] Furthermore, the 1-year mortality after hip fracture was 15% among females and 22% among males.[13] The lifetime risk of vertebral, hip and wrist fractures is 33% and 20% for females and males, respectively.[13] Only one-fourth of Taiwanese patients receive bone mineral density (BMD) testing, and one-third undergo treatment.[14] Since 2014, FLSs have been actively developed throughout Taiwan to improve osteoporosis care.

Secondary Fracture Prevention. https://doi.org/10.1016/B978-0-12-813136-7.00009-0

THE DEVELOPMENT OF OSTEOPOROSIS CARE MANAGEMENT SERVICES IN TAIWAN

Before the introduction of FLSs, the most thorough osteoporosis case management service in Taiwan was operated by Lukang Christian Hospital (LCH). After 6 years of experience, the prescriptions, examinations, evaluations and communications between case managers and physicians were implemented with the help of a comprehensive computer management system. However, the LCH program focused on primary prevention by identifying high-risk patients with low BMD and not on secondary fracture prevention. The latter is the focus of FLSs, which are designed to identify patients with pre-existing or incident fractures. The first medical institute to implement the Best Practice Framework for secondary Fracture Prevention is the National Taiwan University Hospital (NTUH) Healthcare System in 2014. This model serves as the backbone for many other programs in Taiwan.

THE NATIONAL TAIWAN UNIVERSITY HOSPITAL MODEL

The FLSs at NTUH and its affiliated Bei-Hu branch were initially operated as a research project. The NTUH research ethic committee approved the study in 2014. The team included orthopaedists, geriatricians, endocrinologists, family physicians and other specialists who discussed the cooperation and communication strategy in the beginning of the project. After recruiting the case coordinators, they served as a bridge in communicating between patients and physicians as well as among team members. In this closed system, patients were identified, investigated and initiated on treatment within the FLSs program. Potential cases were screened from both inpatients and outpatients although the Bei-Hu branch recruited only outpatients. Fig. 9.1 is a flow chart of the NTUH FLS model.

The NTUH model enrolled new hip fracture cases from its orthopaedic ward (group A), asymptomatic vertebral compression fractures from the geriatrics wards (group B) and patients with symptomatic vertebral compression fractures and no previous treatment from other outpatient clinics (group C). The exclusion criteria were fractures sustained after significant trauma, cancer-related fractures, atypical femoral shaft fractures, enrolment in another clinical intervention trial, life expectancy less than 2 years, inability to undergo evaluation and refusal to participate. New hip fracture cases (group A) were identified through surgical lists in the orthopaedic ward by chief residents; the chief resident then uses instant messaging to inform both attending physicians and coordinators. After a brief introduction by the attending physician, the coordinators obtained informed consent for attending the FLS and performed baseline evaluations. The vertebral fracture cases from the geriatrics ward (group B) were enroled through medical record keyword searches by the case coordinator. When previously untreated vertebral fractures were found (BPF standard 4), coordinators contacted the attending physician who followed the same enrolment process as group A. Patients presented to team physicians' clinics with back pain, loss of

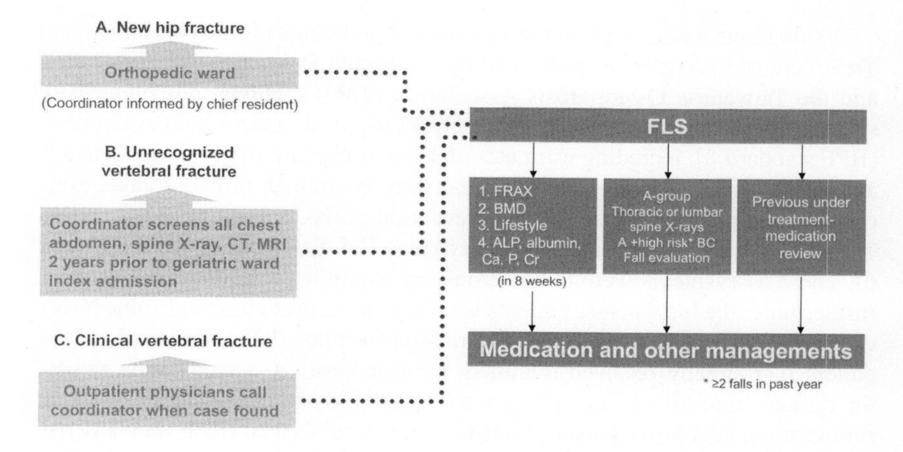

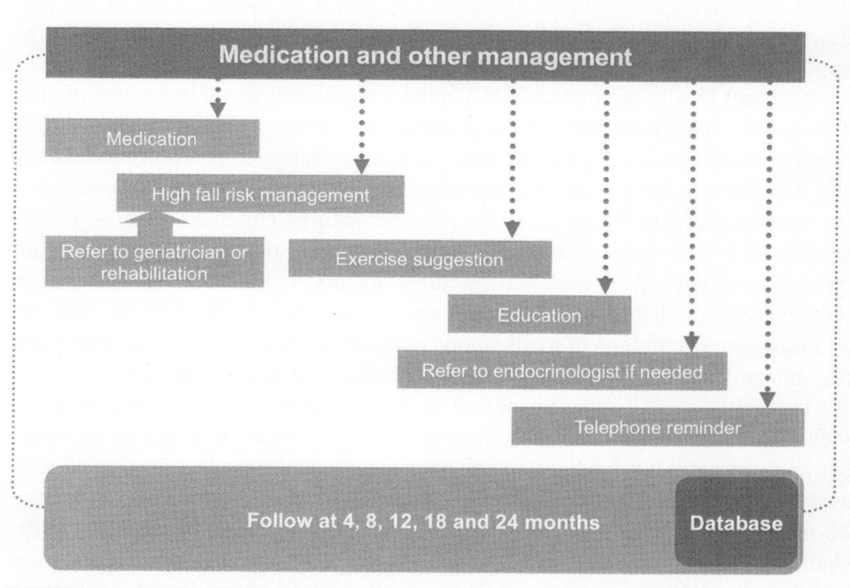

FIGURE 9.1 Flowchart of National Taiwan University Hospital Fracture Liaison Service model. *ALP*, alkaline phosphatase; *BMD*, bone mineral density; *Ca*, calcium; *Cr*, creatinine; *CT*, computed tomography; *FLS*, fracture liaison service; *MRI*, magnetic resonance imaging; *NTUH*, National Taiwan University Hospital; *P*, phosphorus.

body height or other symptoms were examined with spine X-rays to identify compression fractures. The team physicians called coordinators to enrol cases when vertebral fractures were found (group C). At the Bei-Hu branch, one team physician reviewed spine or chest X-rays weekly using the Genant semiquantitative technique.[15] Patients with vertebral fractures were then invited to visit one of the clinics for enrolment and evaluation. The potential cases and enrolment percentages were calculated as well (BPF standard 2).

Evaluations were based on the Taiwanese Guidelines for the Prevention and Treatment of Osteoporosis published by the Health Promotion Administration and the Taiwanese Osteoporosis Association (TAO) (national guideline, BPF standard 5).[13] Patients with fragility fractures received multifaceted evaluations (BPF standard 8), including estimates of 10-year fracture risk through FRAX[16] and BMD tests, lifestyle evaluations (calcium, vitamin D, protein intake, exercise, smoking, etc.), assessment and prevention of risk of falling (BPF standard 7) and blood test for possible secondary causes of osteoporosis (BPF standard 6). These assessments were to be completed within 8 weeks (BPF standard 3); furthermore, the hip fracture patients were also to receive chest and spine X-ray examinations for latent vertebral compression fractures (BPF standard 4). If the patient had already received treatment for osteoporosis, a medication review for dosage, side effects, compliance and contraindications was implemented; furthermore, alternative prescriptions were reviewed as well (BPF standard 10). After enrolment, coordinators provided education on osteoporosis to the participants and updated education on medication use and other information as appropriate.

National Health Insurance (NHI) is the compulsory social insurance plan in Taiwan, covering over 99% of the population.[13] It reimburses only antiresorptive medications (i.e., bisphosphonates, selective oestrogen receptor modulators and denosumab) for patients with osteoporosis (BMD T-score ≤ -2.5) and existing vertebral or hip fractures or for patients with osteopenia (BMD $-1.0 \geq$ T score > -2.5) who present with two or more vertebral or hip fractures. Fracture types other than hip or vertebral fractures are not included in NHI; therefore only patients with hip or spine fractures are enroled in the NTUH FLS. In assessing the percentage of medications initiated among participants, only those reimbursable by the NHI served as the denominator (BPF standard 9).

The initial case follow-up period is 1 year; telephone or outpatient interviews took place after 4, 8 and 12 months. The interviews included patient education, assessment of medication adherence, review of prescription side effects, evaluation of new falls or fractures and new BMD tests if applicable. After the 12-month interview, a long-term osteoporosis care program was organised (BPF standard 12); the follow-up period extended to every 6 months for a second year and then annually up to 10 years. The overall process of enrolment, evaluation and follow-up was consolidated into a flow chart enabling the coordinator to communicate with the physicians (BPF standard 11). Moreover, a FLS database was established to facilitate analysis and further project improvement (BPF standard 13).

PROMOTION OF THE FRACTURE LIAISON SERVICE IN TAIWAN

In 2015, both programs were accredited by the Capture the Fracture as reflecting best practices. Since then, several other medical centres in Taiwan have cooperated with NTUH using similar core protocols. Those collaborating institutions include Linkou Chang Gung Memorial Hospital (LKCGMH), Kaohsiung Medical University Hospital (KMUH) and China Medical University Hospital (CMUH), whereas National Yang-Ming University Hospital (NYMUH) has developed its own system.

Since 2016, the TOA hosted multiple workshops to facilitate the dissemination of FLSs in Taiwan. The workshops invite experienced team members to cooperate and pass along their experience to other institutes trying to establish their own FLSs. As of February 2018, there were 22 secondary fracture prevention programs in Taiwan (Fig. 9.2A), 11 of which have been accredited on the BPF map (Fig. 9.2B; Table 9.1). Therefore, service coverage in Taiwan has been among the best in the Asia-Pacific region. Among the 11 accredited programs, 9 made their BPF questionnaires available to us.

Table 9.1 summarised characteristics of these programs. Hospitals vary significantly in the size of their service populations. In general, tertiary medical centers serve larger populations. Programs also vary significantly in the number

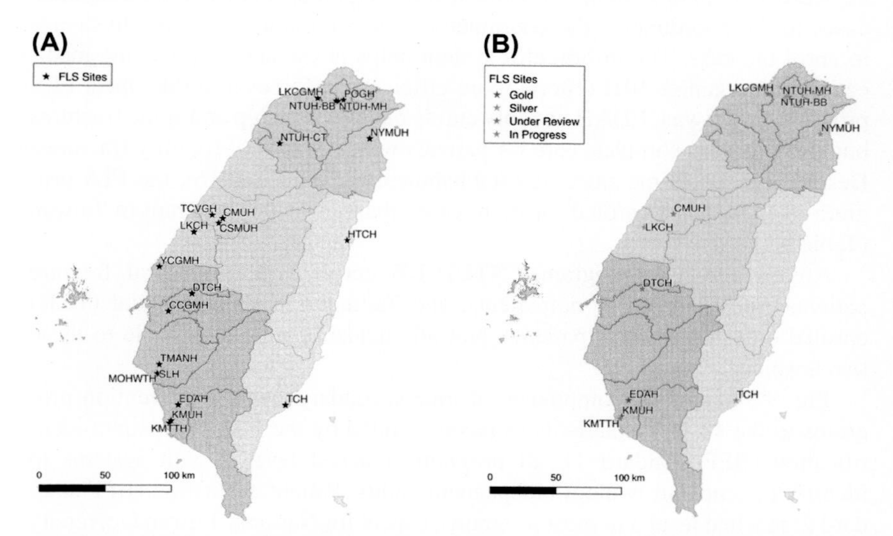

FIGURE 9.2 Fracture Liaison Services Map in Taiwan. (A) Established sites; (B) International Osteoporosis Foundation–accredited sites. *CCGMH*, Chiayi Chang Gung Memorial Hospital; *CMUH*, China Medical University Hospital; *CSMUH*, Chung Shan Medical University Hospital; *DTCH*, Buddhist Dalin Tzu Chi General Hospital; *EDAH*, E-Da Hospital; *HTCH*, Hualien Tzu Chi Hospital; *KMTTH*, Kaohsiung Municipal Ta-Tung Hospital; *KMUH*, Kaohsiung Medical University Chung-Ho Memorial Hospital; *LCGMH*, Linkou Chang Gung Memorial Hospital; *LKCH*, Lukang Christian Hospital; *MOHWTH*, MOHW Tainan Hospital; *NTUH-BB*, National Taiwan University Hospital-Main Hospital Bei-Hu Branch; *NTUH-CT*, National Taiwan University Hospital-Main Hospital Chu-Tung Branch; *NTUH-MH*, National Taiwan University Hospital-Main Hospital; *NYMUH*, National Yang-Ming University Hospital; *POGH*, Pojen General Hospital; *SLH*, Tainan Sin Lau Hospital; *TCH*, Taitung Christian Hospital; *TCVGH*, Taichung Veterans General Hospital; *TMANH*, Tainan Municipal An-Nan Hospital; *YCGMH*, Yunlin Chang Gung Memorial Hospital.

of their enrollees when they apply for best practice recognitions. The numbers were not updated after the application. For example, the FLS program at National Taiwan University Hospital Bei-Hu Branch (NTUH-BB) now serves more than 200 patients. When we summarised data from application forms, there were only 40 patients.

In examining the attributes of the nine programs, several common features can be found. Most FLS programs in Taiwan were established in teaching hospitals, as these services are generally conducted as research programs. However, this characteristic may pose a challenge to nonteaching hospital because of the difficulty involved in acquiring sufficient funding. Most programs enrol only hip or spine fracture patients because they are the only fragility fractures whose osteoporosis medications are reimbursed by the NHI. Lastly, DXA scanners are available at all nine sites, facilitating BMD evaluations and the assessment of future fracture risk.

Among the newly established FLS programs, the E-Da Hospital (EDAH) has performed exceptionally well. Through its computer-assisted diagnostic and keyword searching system, its medical information specialists refer candidate cases to the coordinator; the coordinator then communicates with physicians to enrol the case. The informatics system helps physicians to prescribe medication and organise NHI criteria more efficiently. Moreover, unlike most FLS programs in Taiwan, EDAH not only enrols patients with hip and spine fractures but also provides complete care for patients with all types of fragility fractures. Despite the short time since its establishment (i.e., since 2016), the FLS program in EDAH has enroled far more cases than any other program in Taiwan (Table 9.1).

Among the nine programs, NTUH-BB enroled only vertebral fracture patients from its pool of outpatients, and Taitung Christian Hospital (TCH) enroled only hip fracture patients. Not all standards were applicable to these two hospitals.

Fig. 9.3 visualises compliance of nine secondary fracture prevention programs to the 13 BPF standards as recommended by the IOF. For patient identification (BPF standard 1), all programs reached level 2 with systems to identify patients but without independent audits. Patient evaluation (BPF standard 2) reached level 3 in most programs, except for National Taiwan University Hospital Main Branch (NTUH-MH), Dalin Tzu Chi General Hospital (DTCH), NTUH-BB and China Medical University Hospital (CMUH), with up to 90% of identified patients evaluated by the secondary fracture prevention program. Except for DTCH, the other three programs were research studies, with some patients refusing participation or being excluded from enrolment. Regarding the timing of postfracture assessment (BPF standard 3), all programs completed their assessments by 8 weeks. In Taiwan, DXA availability is generally not an issue in hospitals. With a reasonable waiting time, most assessments can be done within about 1 month. Programs that enrolled both hip and vertebral fractures performed thoracolumbar spine X-ray in patients with hip fracture to identify

TABLE 9.1 Summary of Nine Fracture Liaison Service Programs in Taiwan

Name	Service Population Size	Year FLS Started	Number of Cases at Submission	Institute Type	FLS Coordinator	Patient Identification Restriction	DXA Availability	DXA Restriction
NTUH-MH	c. 8,000,000	2014	407	Teaching/ University	Researcher	≧50 years; hip or vertebral fracture only	Yes	Nontraumatic fracture
DTCH	c. 350,000	2014	655	Private not for profit	Dedicated coordinator	No	Yes	Nontraumatic fracture
NTUH-BB	c. 100,000	2014	40	District	Dedicated coordinator	≧50 years; vertebral fracture only	Yes	Nontraumatic fracture
LCGMH	c. 2,117,000	2015	107	Teaching/ University	Dedicated coordinator	≧50 years; hip or vertebral fracture only	Yes	≧50 y or postmenopausal women; hip or vertebral fracture only
KMUH	c. 300,000	2015	150	Teaching/ University	Researcher	≧50 years; hip or vertebral fracture only	Yes	Nontraumatic fracture
CMUH	c. 100,000	2015	130	Teaching/ University	Dedicated coordinator	≧50 years	Yes	No
EDAH	c. 1,000,000	2016	2849	Teaching/ University	Dedicated coordinator	No	Yes	No

Continued

TABLE 9.1 Summary of Nine Fracture Liaison Service Programs in Taiwan—cont'd

Name	Service Population Size	Year FLS Started	Number of Cases at Submission	Institute Type	FLS Coordinator	Patient Identification Restriction	DXA Availability	DXA Restriction
TCH	c. 223,000	2016	42	Private not for profit	Clinician	≧50 years; hip fracture only	Yes	≧50 y; hip or vertebral fracture only
LCH	c. 75,000	2016	1318	District	Dedicated coordinator	≧50 years; hip or vertebral fracture only	Yes	≧18 y

CMUH, China Medical University Hospital; *DTCH*, Dalin Tzu Chi General Hospital; *EDAH*, E-Da Hospital; *FLA*, Fracture Liaison Services; *KMUH*, Kaohsiung Medical University Memorial Hospital; *LCGMH*, Linkou Chang Gung Memorial Hospital; *LCH*, Lukang Christian Hospital; *NTUH-BB*, National Taiwan University Hospital Bei-Hu Branch; *NTUH-MH*, National Taiwan University Hospital Main Branch; *TCH*, Taitung Christian Hospital.

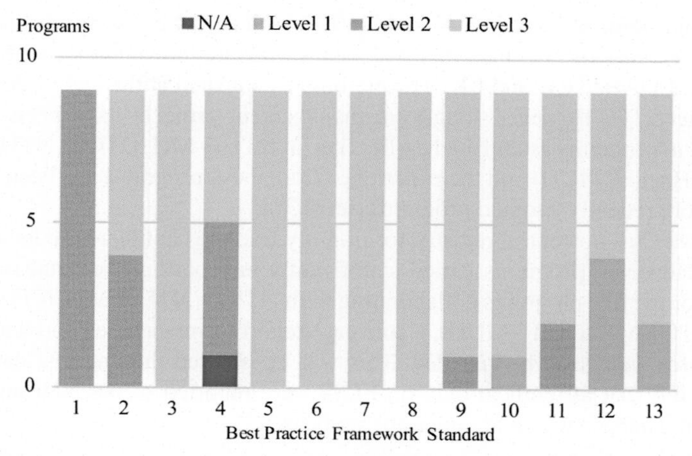

FIGURE 9.3 Compliance of the nine Fracture Liaison Services Programs in Taiwan to the 13 Best Practice Framework Standards.

additional vertebral fractures (BPF standard 4). Most level 3 hospitals conduct keyword searches on image reports either by coordinators or by the radiology information system. NTUH BB assigned a physician to read plain X-rays weekly to identify vertebral fractures. This standard is not applicable to TCH.

All programs reached level 3 for BPF standard 5 by applying the Taiwanese Osteoporosis Practice Guideline.[13,17] Programs varied in their workup for secondary causes for osteoporosis. However, blood testing of serum calcium, albumin (to adjust calcium) phosphate, alkaline phosphate and creatinine (renal function evaluation) was generally done. All programs reached level 3 for BPF standard 6. Official fall prevention services are generally not available in Taiwan. However, all programs identified patients at high risk for falls (e.g., at least two falls in the past year) and provided some basic assessments in fall risk factors, with referral to physiatrists or geriatricians where appropriate. All programs reached level 3 for BPF standard 7. In the baseline assessments, multifaceted health and lifestyle risk factor assessments were included in all programs. Therefore all programs reached level 3 for BPF standard 8.

When only those fractures reimbursable by NHI were counted, most programs (except for Lukang Christian Hospital (LKCH)) managed to initiate medication for 90% or more of the enrolled patients (level 3). If all patients were considered, most programs could still initiate medication for at least 70% (level 2) of participants (BPF standard 9). For BPF standard 10, except for LKCH, all other programs provided medication review if patients had already been treated before enrolment (level 3). Vertebral fracture assessment by DXA or spine X-ray result and fall risk factors is not routinely included in the report of DTCH and TCH; therefore the criteria for communication strategy (BPF standard 11) only reached level 2 for these two hospitals.

A long-term management schedule (BPF standard 12) has not been established because of the fact that secondary fracture prevention programs at CMUH, EDAH, TCH and LKCH were only established at the end of 2015 and 2016 (level 1). However, long-term management projects have been implemented in programs established earlier (i.e., at NTUH-MH, DTCH, NTUH-BB, LCGMH and KMUH) and the experience from these programs has been shared with more recently adopted programs (level 3).

Taiwan has no central registry for fragility fractures but there are local databases. However, programs varied significantly in reporting the database standard. For the five programs collaborating with NTUH (NTUH-MH, NTUH-BB, LCGMH, KMUH and CMUH), a core case report form was used and data were shared for analysis when needed. This was considered the national database. Others felt that all clinical data on FLSs were reported to the NHI and NHI research database which are open for analysis.

FRACTURE LIAISON SERVICES IN TAIWAN: PRELIMINARY OUTCOMES

As mentioned above, most programs in Taiwan reached very high (>90%) BMD testing and medication initiation rates. The 1-year mortality of hip fracture was 16% overall in Taiwan.[18] By analysing the database of NTUH and cooperating hospitals, we found that the new hip fracture patients (group A) had the highest mortality. However, the mortality was about 9% for hip fracture patients enrolled in a FLS, significantly lower than the national average. In particular, the 1-year mortality of KMUH was less than 5%, possibly due to the thorough cooperation between orthopaedic and geriatric physicians (Fig. 9.4).[19] Furthermore, the preliminary analysis of NTUH has shown very high medication adherence (>90%) at 1 year. There were significantly improved (1) calcium and vitamin D supplement intake, (2) percentage of adequate protein intake (>65 g/d) and (3) exercise at 1 year than in baseline data. The 1-year fall rate was also considerably lower than at baseline.

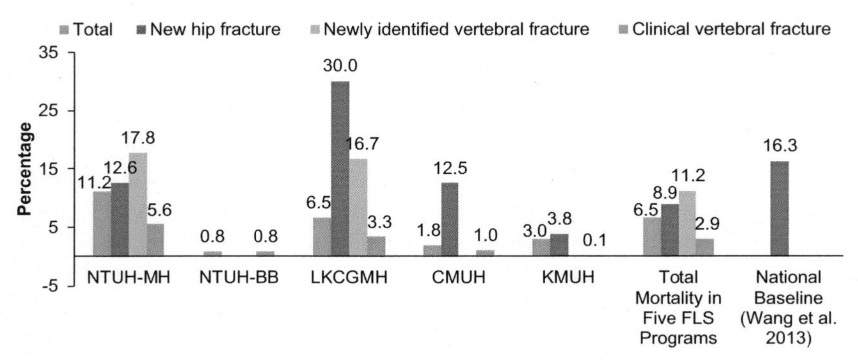

FIGURE 9.4 One-year mortality of five Collaborating Fracture Liaison Service Programs in Taiwan.

CHALLENGES IN ESTABLISHING FRACTURE LIAISON SERVICES IN TAIWAN

Lack of funding is the major challenge in establishing FLSs worldwide, including Taiwan. Some programs receive research funding from philanthropists and from pharmaceutical companies. Most programs reallocate hospital resources to conduct their FLSs. The salaries of coordinators constitute the major costs. With careful calculations, programs can identify potential savings or extra revenues from FLSs in their business plan to persuade higher administrators to provide funding. The EDAH program is a successful model that quickly expanded to sustained funding from the hospital. The unequal distribution of FLSs across Taiwan is shown in Fig. 9.2; most of the programs are located in Northern and Southern Taiwan, which are major metropolitan areas. Patients from Eastern Taiwan and outlying islands have relatively lower access to FLSs.

In conclusion, the secondary fracture prevention programs/FLSs in Taiwan are thriving and are associated with beneficial outcomes. However, randomised studies are still warranted to confirm their effectiveness. Meanwhile, negotiations with the Department of Health and Welfare are under way to fund several pilot studies on FLSs with the goal of establishing a coordinator payment scheme by the NHI in the future. With the rapid ageing of Taiwan's population, FLSs may help to remedy the extraordinary burden of osteoporotic fragility fracture, improve postfracture care and ensure a better quality of life for patients with osteoporosis.

REFERENCES

1. Akesson K, Marsh D, Mitchell PJ, et al. Capture the Fracture: a Best Practice Framework and global campaign to break the fragility fracture cycle. *Osteoporos Int* 2013;**24**(8):2135–52.
2. McLellan AR, Gallacher SJ, Fraser M, McQuillian C. The fracture liaison service: success of a program for the evaluation and management of patients with osteoporotic fracture. *Osteoporos Int* 2003;**14**(12):1028–34.
3. Dell R. Fracture prevention in Kaiser Permanente Southern California. *Osteoporos Int* 2011;**22**(Suppl. 3):457–60.
4. Eisman JA, Bogoch ER, Dell R, et al. Making the first fracture the last fracture: ASBMR task force report on secondary fracture prevention. *J Bone Miner Res* 2012;**27**(10):2039–46.
5. Boudou L, Gerbay B, Chopin F, Ollagnier E, Collet P, Thomas T. Management of osteoporosis in fracture liaison service associated with long-term adherence to treatment. *Osteoporos Int* 2011;**22**(7):2099–106.
6. Wallace I, Callachand F, Elliott J, Gardiner P. An evaluation of an enhanced fracture liaison service as the optimal model for secondary prevention of osteoporosis. *JRSM Short Rep* 2011;**2**(2):8.
7. Wu CH, Tu ST, Chang YF, et al. Fracture liaison services improve outcomes of patients with osteoporosis-related fractures: a systematic literature review and meta-analysis. *Bone* 2018;**111**:92–100.
8. Briot K. Fracture liaison services. *Curr Opin Rheumatol* 2017;**29**(4):416–21.
9. Mitchell PJ. Best practices in secondary fracture prevention: fracture liaison services. *Curr Osteoporos Rep* 2013;**11**(1):52–60.

10. Cooper MS, Palmer AJ, Seibel MJ. Cost-effectiveness of the Concord Minimal Trauma Fracture Liaison service, a prospective, controlled fracture prevention study. *Osteoporos Int* 2012;**23**(1):97–107.
11. Wu CH, Kao IJ, Hung WC, et al. Economic impact and cost-effectiveness of fracture liaison services: a systematic review of the literature. *Osteoporos Int* 2018;**29**:1227–42.
12. Kanis JA, Oden A, McCloskey EV, et al. A systematic review of hip fracture incidence and probability of fracture worldwide. *Osteoporos Int* 2012;**23**(9):2239–56.
13. Health Promotion Administration NHRI, The Taiwanese Osteoporosis Association. *Taiwan osteoporosis practice guidelines*. Taipei, Taiwan: Health Promotion Administration; 2015.
14. Kung AW, Fan T, Xu L, et al. Factors influencing diagnosis and treatment of osteoporosis after a fragility fracture among postmenopausal women in Asian countries: a retrospective study. *BMC Womens Health* 2013;**13**(7).
15. Genant HK, Wu CY, van Kuijk C, Nevitt MC. Vertebral fracture assessment using a semiquantitative technique. *J Bone Miner Res* 1993;**8**(9):1137–48.
16. Kanis JA, Johnell O, Oden A, Johansson H, McCloskey E. FRAX and the assessment of fracture probability in men and women from the UK. *Osteoporos Int* 2008;**19**(4):385–97.
17. Soong YK, Tsai KS, Yang RS, Wu CH, Hwang JS, Chan DC. *Taiwanese guidelines for the prevention and treatment of osteoporosis*. Taipei, Taiwan: The Taiwanese Osteoporosis Association; 2013.
18. Wang CB, Lin CF, Liang WM, et al. Excess mortality after hip fracture among the elderly in Taiwan: a nationwide population-based cohort study. *Bone* 2013;**56**(1):147–53.
19. Chen C-H, Wang H-Y, Huang H-T, et al. The high volume hip surgeon acting as ortho-geriatric reduced mortality in elderly hip fracture after surgery. *Osteoporos Sarcopenia* 2015;**1**(2):142.

Chapter 10

International Models of Secondary Fracture Prevention: United Kingdom

Muhammad Kassim Javaid[1], Paul J. Mitchell[2,3]
[1]National Institute of Health Research Oxford Biomedical Research Centre, Nuffield Department of Orthopaedics, Rheumatology and Musculoskeletal Sciences, University of Oxford, Oxford, United Kingdom; [2]School of Medicine, Sydney Campus, University of Notre Dame Australia, Sydney, NSW, Australia; [3]Osteoporosis New Zealand, Wellington, New Zealand

THE UK NATIONAL HEALTH SERVICE

The National Health Service in the United Kingdom was created on 5 July 1948 by the UK government under the stewardship of Aneurin Bevan. The NHS is based on three principles: that it meets the needs of everyone, that it be free at the point of delivery and that it be based on clinical need, not the ability to pay. Since its creation, the NHS has split into separate entities to cover each of the four nations, England, Scotland, Wales and Northern Ireland. Although recognised as one of the most efficient health-care systems globally, the NHS has a culture of regular and significant reorganisation based on political priorities. This affects the clinical and economic framework within which secondary fracture prevention has to operate. At its core, there is an extensive primary care service led by independent general practitioners (GP) contracted by the NHS to provide routine community-based health care. A component of GPs income is based on demonstration of the quality of care in prespecified clinical areas. The next layer is secondary care that typically provides hospital-based services funded mainly from contracts with regional boards (clinical commissioning groups in England and health boards in Scotland, Wales and Northern Ireland). The aim of these regional boards is to prioritise services based on local not national perceived need within a finite budget. Although newer models are being tested, such as integrated care systems, this is the culture of reorganisation that Fracture Liaison Services (FLSs) need to operate within.

Secondary Fracture Prevention. https://doi.org/10.1016/B978-0-12-813136-7.00010-7

IT STARTED IN GLASGOW, SCOTLAND

In the 1990s, with the advent of clinically effective antiosteoporosis therapies that reduced fracture risk, osteoporosis services were focused on primary prevention based on dual-energy X-ray absorptiometry (DXA) assessment. Against this trend, two endocrinologists, A McClellan and S Gallacher, realised the need to prioritise secondary fracture prevention and ensure those presenting with a fragility fracture had a comprehensive bone assessment. Recognising the scale of the problem, it was clear that a nonphysician-delivered service would be required and the FLS concept was created in 1999. This service was delivered by specialist nurses and importantly was not simply a coordinator role. FLS nurses identified the patients from clinic and wards to a one-stop clinic for risk factor assessment, DXA scanning, laboratory testing and treatment recommendation. Further monitoring was delegated to the patient's primary care physician. This was tremendously successful and led to a number of important publications related to the characterisation of patients[1–4] including one of the first estimates of cost-effectiveness.[5] Realising the importance of this model of care, a system of 'preceptorships' was created, giving interested clinicians within the United Kingdom and further afield, a series of hands-on workshops to understand the rationale, practicalities and experiences of running an FLS. Attendees from these workshops then seeded new FLSs within the United Kingdom and globally, and the concept of the FLS as a service was realised. A key component of the strategy to establish improved secondary fracture prevention within the United Kingdom was to follow a top-down and bottom-up approach.

THE UK TOP-DOWN APPROACH

Realising the need for centralised funding to support the initiation and sustainability of FLS, FLS champions in the United Kingdom, including senior orthopaedic surgeons, worked with the Department of Health to produce a series of documents focusing on secondary fracture prevention.[6] This culminated with a health economic model, based on a population of 300,000 for FLS[7] and the publication of the 'pyramid' of fracture prevention that uniquely placed hip fracture care, secondary fracture prevention, falls reduction and healthy ageing into a single, simple operational framework (see Fig. 10.1).

Building on the success of the National Hip Fracture Database (NHFD) in driving quality of care for patients, D Marsh led an initiative to pilot and then introduce an audit of health-care outcomes within FLS. An important aspect to the success of this audit has been the multidisciplinary constituency of the audit's advisory group including patients and carers, patient societies, primary carers, rheumatologists, endocrinologists, orthopaedic surgeons, physiotherapists, statisticians and audit methodologists. Each constituent member has an equal voice and outputs are codeveloped. Key deliverables from this pilot were the estimation of the local FLS estimated caseload by multiplying the number of hip fractures,

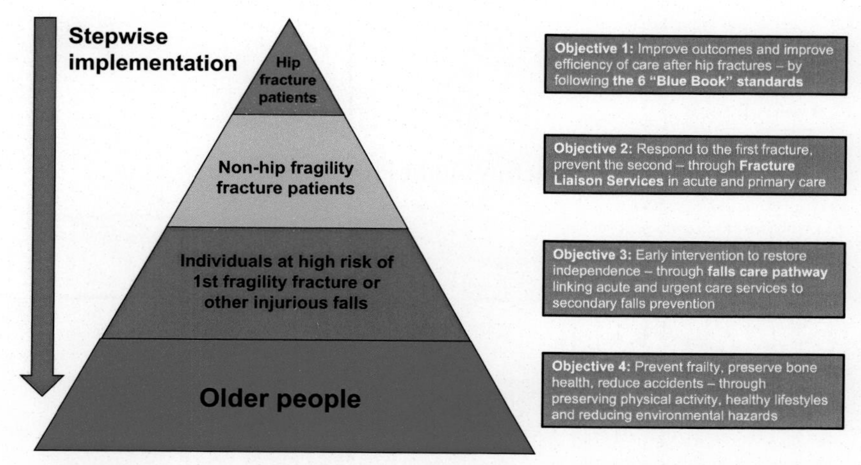

FIGURE 10.1 Department of Health systematic approach to falls and fracture prevention.[6]

the rule of five[8] and the realisation of the information governance barriers and variable data quality of routine primary care data. From this, the Health Quality Improvement Partnership–funded national FLS Database audit was launched in 2016. Although the importance of monitoring adherence to therapy was recognised, given the governance issues, it was decided to delegate reporting of adherence to the FLSs and not develop linkage to primary care data. The FLS Database Audit has since published three reports covering facilities and patient level FLS outcomes.[9–11] The reports allow FLS to compare outcomes and, to seek and learn from those services who are achieving better results for their patients. Key outputs from these reports were (1) recognition of the almost total mismatch between staff numbers paid for within an FLS and the volume of patients that were needed to be seen; (2) development of a set of national key performance indicators for an FLS; (3) recognition that key indicators (e.g., patients with spine fractures and treatment initiation by 16 weeks and adherence at 12 months) are a national priority for quality improvement. The NHFD, FLS Database and the National Audit of Inpatient Falls collectively constitute the Falls and Fragility Fracture Audit Programme managed by the Royal College of Physicians (see Fig. 10.2).[12]

In parallel, the UK National Osteoporosis Society (NOS) has been a tremendously effective strategic player in this area. Activities include the All Party Parliamentary Osteoporosis Group of members of parliament and peers with an interest in osteoporosis; the FLS implementation group, a high level strategy group bringing together the stakeholders in FLS such as NHS England, Public Health England, Royal College of Physicians and representatives from the Scotland, Wales and Northern Ireland that served to coordinate the activities from policy makers to payers; developing the free to use online FLS Benefits and cost calculator that allows payers at the national and regional level, understand the cost savings and the costs associated with the introduction of an FLS

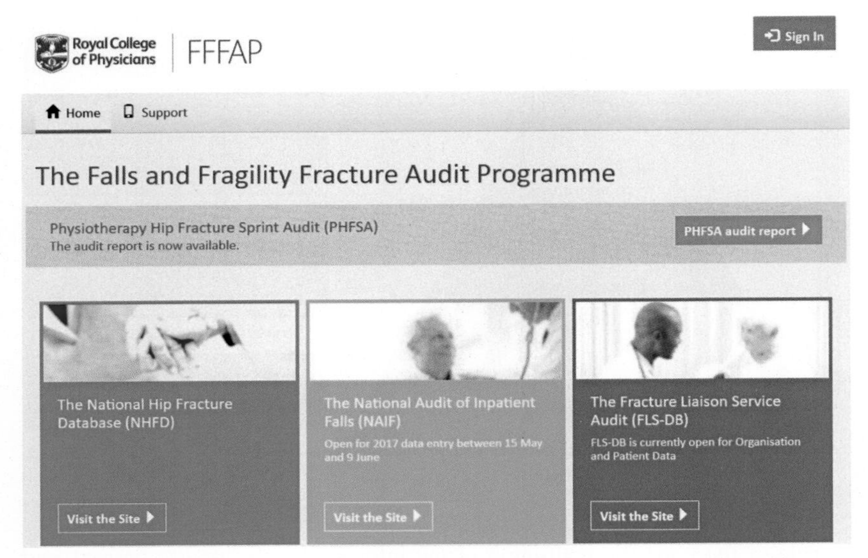

FIGURE 10.2 The Royal College of Physicians Falls and Fragility Fracture Audit Programme.[12]

(https://benefits.nos.org.uk/) and quality standards for a FLS[13] and vertebral fractures in addition to the Best Practice Framework from the International Osteoporosis Foundation.[14]

Since 2005, the National Institute for Health and Care Excellence (NICE) has developed a comprehensive suite of technology appraisals, clinical guidelines and quality standards relating to the management and prevention of fragility fractures:

- NICE Technology Appraisals (TA):
  - NICE TA 464 (update): Bisphosphonates for treating osteoporosis (updated February 2018).[15]
  - NICE TA 161 (update): Raloxifene and teriparatide for the secondary prevention of osteoporotic fragility fractures in postmenopausal women (updated February 2018).[16]
  - NICE TA 160 (update): Raloxifene for the primary prevention of osteoporotic fragility fractures in postmenopausal women (updated February 2018).[17]
  - NICE TA 204: Denosumab for the prevention of osteoporotic fractures in postmenopausal women.[18]
- NICE Clinical Guidelines (CG):
  - NICE CG 146 (update): Osteoporosis: assessing the risk of fragility fracture (updated February 2017).[19]
  - NICE CG 124 (update): Hip fracture: management (updated May 2017).[20]
- NICE Quality Standards (QS):
  - NICE QS 149: Osteoporosis.[21]
  - NICE QS 16 (update): Hip fracture in adults (updated May 2017).[22]

To support implementation of the policies and guidelines described above, financial incentives linked to delivery of best practice have also been established in the United Kingdom during the last decade. In 2010, the Best Practice Tariff (BPT) for hip fracture care was introduced by the Department of Health in England.[23] The BPT offered an uplift in reimbursement for provision of hip fracture care at the individual patient level (made possible by the NHFD). The payment differential for delivering best practice was initially set at £445 for 2010–11, which was subsequently increased to £890 for 2011–12 and £1335 for 2012–13[24] and thereafter.[25] To receive the BPT uplift, all of the following criteria are needed to be met during 2010–11 and 2011–12:

- Time to surgery within 36 h from arrival in an emergency department, or time of diagnosis if an inpatient, to the start of anaesthesia.
- Involvement of an (ortho-) geriatrician:
  - Admitted under the joint care of a consultant geriatrician and a consultant orthopaedic surgeon.
  - Admitted using an assessment protocol agreed by geriatric medicine, orthopaedic surgery and anaesthesia.
  - Assessed by a geriatrician (as defined by a consultant, nonconsultant career grade, or specialist trainee ST3+) in the perioperative period (defined as within 72 h of admission).
  - Postoperative geriatrician-directed:
    - Multiprofessional rehabilitation team.
    - Fracture prevention assessments (falls and bone health).

From April 2012, an additional BPT criterion was added which required pre- and postoperative cognitive assessments to be completed.[24]

In parallel to the hospital-based BPT, in 2012, secondary fracture prevention was included in the General Practice Quality and Outcomes Framework (QOF).[26] The QOF is a system for performance management and payment of GPs in the NHS throughout the United Kingdom. Unfortunately, the limited reimbursement from this system did not have the desired impact in delivering systematic secondary fracture prevention, and it is estimated from the QOF returns that only 25%–30% of patients with a fragility fracture are currently captured within this framework. Another pilot using enhanced reimbursement for GPs within a region has also not been able to deliver systematic secondary fracture prevention to their population.[9] This suggest that purely reimbursement-based methods to improve secondary fracture prevention in primary care may not be sufficient, and further evidence of the effectiveness of primary care-based models of FLS are needed.

THE UK BOTTOM-UP APPROACH

One of the biggest barriers for the FLS is that the patient's pathway crosses departments in secondary care (orthopaedics and trauma, orthogeriatrics, rheumatology, endocrinology, radiology) as well as into primary care (for treatment prescription and monitoring). The publication of the *Effective interventions in health and social care*[27] was followed by a series of multidisciplinary regional workshops, funded by the Department of Health, with the aim of supporting the implementation of the recommendations. Strategically, these workshops brought together local experts, including orthopaedic surgeons and physicians, to work together on local solutions for getting an FLS funded. The Royal College of Physicians, through their audit work, developed a national FLS champions network, that was then handed over to the NOS and this provided a national forum for FLS to share experiences and upskill in key areas. The NOS has proved an essential driver for local FLS development. They have achieved this through a talented Service Delivery team with experience and expertise in running an FLS and commissioning covering all four nations. The NOS Service Delivery team provides toolkits, presentations and 1:1 mentoring for FLS to get commissioned and started. The FLS Benefits calculator (https://benefits.nos.org.uk/) is used to define the local costs and savings from an FLS, and local FLS development workshops take groups of clinicians from local prioritisation to funding applications and service design, as well as focused workshops on aspects of FLS delivery at their national conference.

CURRENT CHALLENGES – DELIVERING FOR EVERY PATIENT

The development of secondary fracture in the NHS that meets the needs of everyone and based on clinical need poses a number challenges including:

1. The identification of patients with spine fractures. Traditional models have focused on patients presenting to spine clinics with a fracture. This is recognised as only the tip of the iceberg and the annual incidence of spine fractures may approximate to the local number of hip fractures, and in the first few years of an FLS identifying all radiographic vertebral fractures, the numbers identified could equal the number of all other fractures combined. Effective and efficient ways to detect patients with a spine fracture are being evaluated in the NHS, and we await the real-world findings.
2. Integration of effective and efficient falls interventions. Bone therapy is only one aspect of fracture prevention. Intervening to reduce falls risk is also important, and there are a number of interventions from the Otago strength and balance programme, medication review and home occupational therapy review. Real-world evidence of the types service models to do this effectively is needed, especially for delivering rapid initiation of falls prevention interventions and longer-term adherence to them to often frail patients.
3. FLS real-world effectiveness. In 2016, Mitchell et al. reviewed models of care for secondary fracture prevention and noted that FLS are associated with significant reductions in refracture rates and mortality.[28] However, unless an FLS can demonstrate, at the patient level, high rates of identification across fracture sites, ages and levels of frailty, systematic assessment to target those

needing bone and falls reduction therapy, initiation of interventions early after fracture and adherence to those interventions, then any observed health or economic benefit is unlikely to be causal[29] and not scalable. A natural experiment, from the Oxford group, looked at fractures after a hip fracture before and after the introduction of an FLS in 11 hospitals and demonstrated no reduction in refracture rates.[30] This is supported by qualitative work where FLS practitioners rated monitoring of therapy adherence as the single highest priority for an FLS, at a time when none of the FLS in the analysis was resourced to include monitoring.[31] This challenges FLSs using the one-stop clinic or coordinator-based models of care in the NHS. The implementation of the patient level FLS audit collecting key clinical data that identifies the effectiveness of an FLS but also supports an FLS to prioritise which components of their pathways to focus on for quality improvement is now in place.[10]

4. Quality improvement: Another major challenge for FLS is to review their local data and implement effective, efficient service improvements that also enhance patient experience within scarce health-care resources. It relies on FLS having a transparent integrated system of self-improvement and to expect to change their working pathways given the ever-changing NHS and society. This will build resilience and sustainability of FLS. FLS working between secondary and primary care is one approach. Operationalising this for every patient in every hospital poses significant capacity issues, including availability of trained FLS nurses. Another approach is to move more of the FLS activity to patient self-management. This will require a different approach for meaningful patient codevelopment of pathways, co-decision-making and co-responsibility. It will need training of FLS teams on how to work effectively with patients as equal partner in service delivery as well as patients as recipients of the FLS. The Fracture Liaison Service Database Audit provides a working structure to benchmark this, and other new FLS models of case within the NHS, to ensure secondary fracture prevention remains effective, efficient and delivers a good patient experience for everyone as a basic NHS provision.

ACKNOWLEDGMENTS

MK Javaid would like to acknowledge the fracture prevention practitioners, administrators, programmers and pharmacists at the Oxford University Hospital Foundation who have worked with him over the last 8 years to improve patients outcomes after a fragility fracture; and also the team at the Royal College of Physicians Falls and Fragility Fracture Audit Programme, the team at the National Osteoporosis Society, the team at the International Osteoporosis Foundation, Capture the Fracture working group and many FLS champions and patients in the United Kingdom who have committed to working together to close the care gap.

DISCLOSURE

MK Javaid (MKJ): In last 5 years, MKJ received honoraria, unrestricted research grants, travel and/or subsistence expenses from Amgen, Eli Lilly, Shire, Internis, Consilient Health, Stirling

Anglia Pharmaceuticals, Mereo Biopharma, Optasia, Zebra Medical Vision, Kyowa Kirin Hakin, UCB.

PJ Mitchell (PJM): Since 2005, PJM has undertaken consultancy for governments, national osteoporosis societies, health-care professional organisations and private sector companies relating to systematic approaches to fragility fracture care and prevention.

REFERENCES

1. McLellan A, Reid DM, Forbes K, Reid R, Campbell C, Gregori A, et al. *Effectiveness of strategies for the secondary prevention of osteoporotic fractures in Scotland: CEPS: 99/03. NHS quality improvement Scotland.* 2004.
2. Gallacher SJ. Setting up an osteoporosis fracture liaison service: background and potential outcomes. *Best Pract Res Clin Rheumatol* 2005;**19**(6):1081–94.
3. Langridge CR, McQuillian C, Watson WS, Walker B, Mitchell L, Gallacher SJ. Refracture following fracture liaison service assessment illustrates the requirement for integrated falls and fracture services. *Calcif Tissue Int* 2007;**81**(2):85–91.
4. McLellan AR, Gallacher SJ, Fraser M, McQuillian C. The fracture liaison service: success of a program for the evaluation and management of patients with osteoporotic fracture. *Osteoporos Int* 2003;**14**(12):1028–34.
5. McLellan AR, Wolowacz SE, Zimovetz EA, Beard SM, Lock S, McCrink L, et al. Fracture liaison services for the evaluation and management of patients with osteoporotic fracture: a cost-effectiveness evaluation based on data collected over 8 years of service provision. *Osteoporos Int* 2011;**22**(7):2083–98.
6. Department of Health. In: Department of Health, editor. *Falls and fractures: effective interventions in health and social care.* 2009.
7. Department of Health. *Fracture prevention services: an economic evaluation.* 2009.
8. Marsh D, Martin F, Tsang C, Cromwell D, Javaid MK, Boulton C, et al. In: Royal College of Physicians, editor. *Secondary fracture prevention: first steps to a national audit.* 2015.
9. Boulton C, Gallagher C, Rai S, Tsang C, Vasilakis N, Javaid MK, Royal College of Physicians. *Fracture Liaison Service Database (FLS-DB) clinical audit. FLS forward: identifying high-quality care in the NHS for secondary fracture prevention.* London: Royal College of Physicians; 2017.
10. Javaid MK, Boulton C, Gallagher C, Judge A, Vasilakis N. In: Royal College of Physicians, editor. *Fracture Liaison Service Database (FLS-DB) annual report: leading FLS improvement: secondary fracture prevention in the NHS.* London: Healthcare Quality Improvement Partnership; 2017.
11. Javaid MK, Rai S, Schoo R, Stanley R, Vasilakis N, Tsang C. In: *Fracture Liaison Service (FLS) Database facilities audit. FLS breakpoint: opportunities for improving patient care following a fragility fracture.* London: Royal College of Physicians; 2016.
12. Royal College of Physicians. *The falls and fragility fracture audit programme.* 2018. Available from: https://www.fffap.org.uk/FFFAP/landing.nsf/index.html.
13. Gittoes N, McLellan AR, Cooper A, Dockery F, Davenport G, Goodwin V, et al. *Effective secondary prevention of fragility fractures: clinical standards for fracture liaison services.* Camerton: National Osteoporosis Society; 2015.
14. Akesson K, Marsh D, Mitchell PJ, McLellan AR, Stenmark J, Pierroz DD, et al. Capture the fracture: a best practice framework and global campaign to break the fragility fracture cycle. *Osteoporos Int* 2013;**24**(8):2135–52.

15. National Institute for Health and Care Excellence. *Bisphosphonates for treating osteoporosis: technology appraisal guidance [TA464]*. London. 2018.
16. National Institute for Health and Care Excellence. *Raloxifene and teriparatide for the secondary prevention of osteoporotic fragility fractures in postmenopausal women: technology appraisal guidance [TA161]*. London. 2018.
17. National Institute for Health and Care Excellence. *Raloxifene for the primary prevention of osteoporotic fragility fractures in postmenopausal women: technology appraisal guidance [TA160]*. London. 2018.
18. National Institute for Health and Care Excellence. *Denosumab for the prevention of osteoporotic fractures in postmenopausal women: technology appraisal guidance [TA204]*. London. 2010.
19. National Institute for Health and Care Excellence. *Osteoporosis: assessing the risk of fragility fracture: clinical guideline [CG146]*. London. 2017.
20. National Institute for Health and Care Excellence. *Hip fracture: management: clinical guideline [CG124]*. London. 2017.
21. National Institute for Health and Care Excellence. *Osteoporosis: quality standard [QS149]*. London. 2017.
22. National Institute for Health and Care Excellence. *Hip fracture in adults: quality standard [QS16]*. London. 2012.
23. Neuburger J, Currie C, Wakeman R, Tsang C, Plant F, De Stavola B, et al. The impact of a national clinician-led audit initiative on care and mortality after hip fracture in England: an external evaluation using time trends in non-audit data. *Med Care* 2015;**53**(8):686–91.
24. Department of Health. In: Department of Health, editor. *Payment by results guidance for 2012-13*. Leeds; 2012.
25. Payment by Results team. In: Department of Health, editor. *Payment by results guidance for 2013-14*. Leeds; 2013.
26. Marshall M, Roland M. The future of the quality and outcomes framework in England. *BMJ* 2017;**359**:j4681.
27. DH/SC LCdOPaD. In: Department of Health, editor. *Falls and fractures: Effective interventions in health and social care*. 2009.
28. Mitchell P, Akesson K, Chandran M, Cooper C, Ganda K, Schneider M. Implementation of Models of Care for secondary osteoporotic fracture prevention and orthogeriatric Models of Care for osteoporotic hip fracture. *Best Pract Res Clin Rheumatol* 2016;**30**(3):536–58.
29. Levesque LE, Hanley JA, Kezouh A, Suissa S. Problem of immortal time bias in cohort studies: example using statins for preventing progression of diabetes. *BMJ* 2010;**340**:b5087.
30. Hawley S, Javaid MK, Prieto Alhambra D, Arden NK, Lippet J, Cooper C, et al. Clinical effectiveness of orthogeriatric and fracture liaison service models of care for hip fracture patients: controlled longitudinal study. *Osteoporos Int* 2015. OC9 Abstract to ESCEO 2015.
31. Drew S, Gooberman-Hill R, Farmer A, Graham L, Javaid MK, Cooper C, et al. Making the case for a fracture liaison service: a qualitative study of the experiences of clinicians and service managers. *BMC Musculoskelet Disord* 2015;**16**:274.

Chapter 11

Fracture Liaison Service: US Perspective

Thomas P. Olenginski
Department of Rheumatology, Geisinger Health System, Danville, PA, United States

INTRODUCTION

Programmatic postfracture care is critically important in 2018 and an aspect of clinical care that demands effective cooperation between orthopaedic surgeons and a committed medical team. There is no dogmatic 'one size fits all' Fracture Liaison Service (FLS) team prescription to espouse, as this critical team can involve rheumatology, endocrinology, family practice, general internal medicine, geriatrics or a multidisciplinary medical team, often with mid-level providers (Registered Nurse; Certified Registered Nurse Practitioner [CRNP]; Physician's Assistant). Additionally, this team can be orthopaedic-led or medically led, provided appropriate team staffing exists. The FLS model was originally developed in the United Kingdom by McLellan et al.[1,2] and other international leaders.[3–6] As a late addition to the National Bone Health Alliance (NBHA) Secondary Fracture Prevention Committee, the author quickly became aware of the importance of FLS postfracture care. He has helped lead an FLS programme at the Geisinger Health System since 2008, termed Geisinger High-Risk Osteoporosis Clinic (HiROC).[7] The author recently participated in the Fracture Prevention Stakeholder Summit (July 19, 2017, Crystal City, VA), organised by the Center for Medical Technology Policy where a 'Strategic Roadmap to Prevent Secondary Fractures' draft has been presented to American Society for Bone Mineral Research (ASBMR), stimulating a heightened resolve and renewed purpose to contribute to developing better solutions to foster postfracture care in the United States.

The FLS model of care simultaneously evolved in several countries at the turn of the century. This included programmes from Australia,[8] Canada,[3] the United Kingdom[1,2] and the United States.[7,9,10] Importantly, a recent review of FLS programmes clearly points to their international flavour and their experience in such programmatic care.[11] The breadth and significance of osteoporosis clinical care gaps has been well documented.[12–14] While Kaiser Permanente and its Healthy Bones Programme and Geisinger with its branded HiROC

Secondary Fracture Prevention. https://doi.org/10.1016/B978-0-12-813136-7.00011-9

Programme have been instrumental in the American FLS movement, the United States Surgeon General's 2004 Report on Bone Health and Osteoporosis was pivotal in drawing attention to the need for a comprehensive plan and call to action in regards to the problem of osteoporosis, its future predicted epidemiology and societal impact, increasing fractures – with attendant morbidity and mortality, treatment, education, patient engagement, research and a multidisciplinary approach.[15] Despite this historic report and the work involved in its preparation, ongoing postfracture care gaps continue to be documented, now 14 years later.[16–18]

In 2009, the Bone and Joint Decade Global Network Conference was held in Washington, DC. Additionally, the American Orthopaedic Association (AOA) established Own the Bone, a programme designed to associate osteoporosis and the ongoing problem of fragility fractures and their morbidity and mortality. Own the Bone also supported the need for FLS programmes.[17] In 2010, the NBHA was founded and its efforts in collaboration with the National Osteoporosis Foundation (NOF), ASBMR, AOA and other key stakeholders have helped to guide the FLS movement in the United States.[18] This cross-sectional network has been invaluable in attempts to raise awareness and support efforts directed to secondary fracture prevention.

In 2012, the NOF and NBHA launched the 2 MILLION 2 MANY national disease awareness campaign, with the unveiling of Cast Mountain, an installation designed to showcase the association between osteoporosis and broken bones.[19] Cast Mountain, a 12 × 12 foot structure (Fig. 11.1), was created to illustrate the daily burden of fracture care in the United States and to challenge health providers to direct efforts to secondary fracture prevention. Since that time, NBHA and NOF have worked cooperatively to enhance this movement, and particularly at NOF's annual Interdisciplinary Symposium on Osteoporosis (ISO) meeting, began to offer an 'FLS Symposium' in efforts to promote expansion of such programmatic and cooperative care between orthopaedics and medical teams. Additionally, similar symposia have been held at other allied professional society meetings. These symposia have helped address making a business case to healthcare leaders and other policymakers and have utilised experienced physicians, nurse practitioners, physician assistants, registered nurses, physical and occupational therapists and other bone team members to illustrate current FLS efforts in US bone healthcare.[18–20] Sharing of best practices continues to be a dominant theme. Despite these efforts, many healthcare leaders, CEO's, and Chief Medical Officers have been slow to fully support FLS program expansion nationwide.

One of the current challenges remains the lack of an FLS registry in the United States (akin to the International Osteoporosis Foundation Capture the Fracture Map of Best Practice]), although NBHA is currently working on a survey that will lead to such a registry of US FLS Programmes.[18,21] The NBHA website provides a suite of resources about FLS, including an archive of webinars designed to address components of FLS care and use shared resources designed to help share best practices. Another challenge continues to be the

FIGURE 11.1 Cast Mountain

cost of FLS programme startup, database registries and programme development. With the support of NBHA and CECity, a demonstration project using a web-based cloud registry was initiated with MedStar Georgetown, University of Pittsburgh Medical Center and Creighton Allegiant Systems.[18,22]

Additionally, NBHA has worked on designing and implementing a 'return on investment calculator' to help assist FLS programmes in working with key leaders, administrators and decision makers in programme startup.[18]

Throughout the United States, all types of FLS programmes exist, though many more need to be developed, mature and persist as key components of postfracture care. It is the intent of this chapter to highlight certain mature US programmes and key leaders in this movement, to summarise key components of their FLS structure, to document data on performance measures in postfracture care and to address ongoing care gaps and proposed solutions and to challenge health-care leaders and policymakers to support the FLS movement in America.

Kaiser Permanente Healthy Bones Programme

In 1945, this integrated managed care network was founded in California. It consists of three entities: (1) Kaiser Foundation Health Plans; (2) Kaiser

Foundation Hospitals and (3) regional Permanente medical groups.[23] Currently, Kaiser operates in eight states plus the District of Columbia and is the largest managed care organisation in America.[23,24]

Kaiser is a single-payor system responsible only for Kaiser Health Plan patients. Kaiser has a fully integrated electronic health-care technology system. Kaiser is committed to quality and innovation and is respectful of health-care costs, health-care expenditures and health-care savings. Their osteoporosis management programme is a blend of case finding (identification of fractured patients), identification of patient groups targeted as 'at risk' (fractures where dual-energy X-ray absorptiometry [DXA] has not been performed; hip fracture patients not on treatment; DXA T-score osteoporosis not on treatment) and in-time identified care gaps (women >60 without DXA; men >70 without DXA; previously treated patients found to have stopped obtaining medication refills). To emphasise, Kaiser's Healthy Bones Programme identifies and narrows in-time care gaps because of their registry's software capabilities. Additionally, Kaiser has been able to determine cost of fractures, especially hip fracture, determine the cost of yearly programme expenditures and show real cost savings in a system committed to fiscal responsibility.[23–25]

In 2011, Dell et al. outlined Kaiser's efforts at fracture prevention.[9] The start of their programme can be traced to 1998 and then, in 2002, the Healthy Bones Programme was staffed at all 11 'regional medical centres' in Southern California. Dell and Kaiser began with their 'Simple Model'[9]:

S – simple in design
I – inexpensive to start and maintain
M – measurable
P – pays for self
L – lasts
E – evolves

Dell emphasises Kaiser's continuous quality improvement (CQI) culture, its innovative medical information technology, clearly a key component of the programme, and finally, that 'best ideas come from dedicated frontline clinicians'. Dell also acknowledges that Kaiser has learned from programmes developed in Glasgow and Geisinger and emphasises use of Plan-Do-Study-Act (PDSA) cycles in their CQI philosophy.

In enumerating their 10 steps, Dell signifies the importance of 'work lists'; identification of care gaps as paramount on these lists; importance of ongoing measurements and incentives to all groups involved in the work.

Steps 1 and 2 begin with planning and goal setting. Steps 3 and 4 emphasise 'case finding' through the information system, namely identifying hip fracture patients and those older than 70 years with other fractures, where treatment and evaluation are indicated. Step 5 invests in midlevel providers – nurses and nurse practitioners, providing oversight of 'work lists' and serving as Care Managers – identifying newly fractured patients; patients with new DXAs and an osteoporosis diagnosis; women >65 years and men >70 years who have never had DXA and those previously on osteoporosis prescription medication, where failure to refill medication is electronically identified. This Patient Management Tool is available to all bone team members and allows Case Managers to solve gaps in real time. Step 6 involves measurements, but unique to Kaiser, that gives any physician champion the ownership to analyse, strategise and capitalise, whereas the Care Managers are working the lists and providing the care, with the consent of primary care physicians. Step 7 involves checking for variation in care patterns. Step 8 emphasises a 'carrot over stick' approach to incentive versus punitive rules. Finally, steps 9 and 10 emphasise goal setting and challenging one's team.[9] Figs. 11.2–11.4 illustrate and document the effectiveness of the Kaiser Healthy Bones Programme, with demonstrated increase in DXA utilisation, osteoporosis treatment and reduction in hip fractures.

Over this time, the osteoporosis management in older women after fracture Healthcare Effectiveness Data and Information Set (HEDIS) measurement has been reported across all Medicare health plans. To meet this metric, any woman with fracture after age 67 years must be placed on osteoporosis prescription medication or have a DXA scan within 6 months of fracture. In 2010, the average HEDIS measure in the United States was about 20%. Kaiser

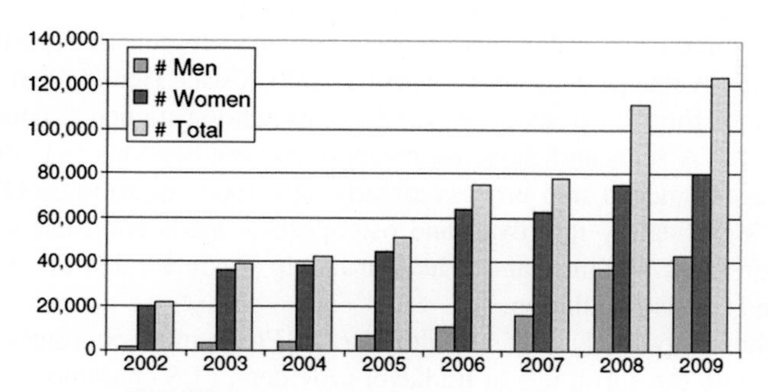

FIGURE 11.2 Kaiser DXA Utilisation Increases

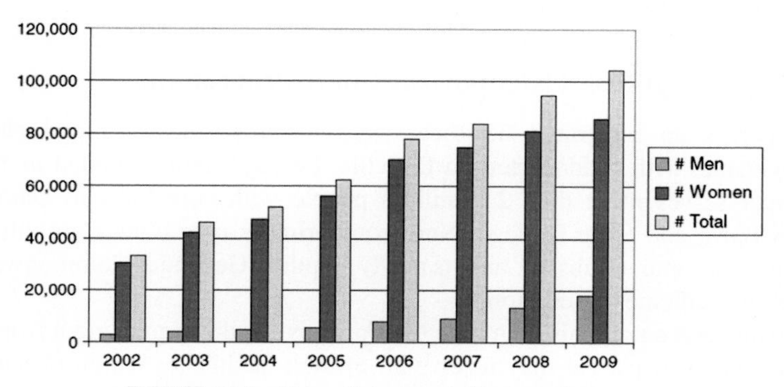

FIGURE 11.3 Kaiser Osteoporosis Treatment Increases

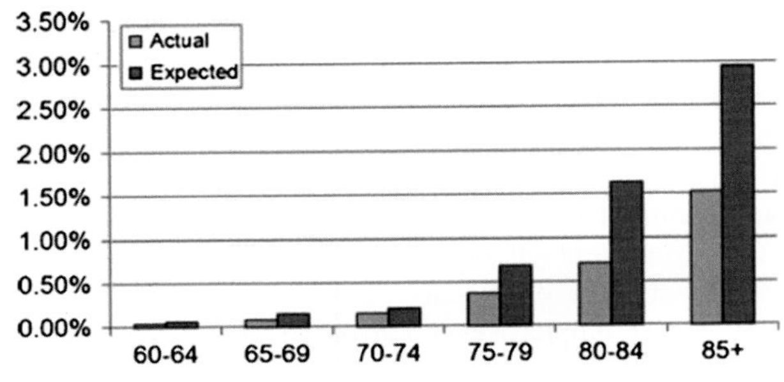

FIGURE 11.4 Kaiser Hip Fracture Rates Reduced

Healthcare plans have consistently led American health plans in this measure, their 2016 measurement for osteoporosis management in women after a fracture being 90%.[23–26]

Kaiser's investment in midlevel providers (nurses, nurse practitioners, physician assistants) has yielded dividends in performance measurement success and health-care savings. Their role as Care Managers, ability to work lists, identify gaps with their 'Just-in-Time' approaches in 2010 cost $5 million US dollars but yielded a 10-fold return in healthcare savings.[23–26]

Finally, Greene et al.'s 2010 report similarly emphasises performance measure improvements in a study with 650,000 Kaiser Health Plan patients. Included were those >60 years; those >50 years who sustained fracture; those who had a DXA scan and those on osteoporosis medications.[27] Utilising the Care Managers and 10 step process already described, appropriate DXA scan utilisation significantly improved and osteoporosis medication use increased. Most importantly, a significant reduction in hip fracture rates of 38.1% was documented in this population.[27]

Since these reports, Kaiser's Healthy Bones Programme continues and sustains itself with the adroit use of midlevel providers, FLS champions, a system commitment to their Healthy Bones Programme and ongoing measurements and quality assurance.

Geisinger High-Risk Osteoporosis Clinic Programme

Geisinger is an integrated healthcare system in northeastern and central Pennsylvania, with headquarters in Danville, Pennsylvania. Founded in 1915, Geisinger serves more than 2.6 million people with core hubs in Danville, Wilkes-Barre and State College. Numerous primary care clinics are situated between these major hubs[28,29] and its newly acquired Geisinger Commonwealth School of Medicine in Scranton.

Geisinger is an organisation with multiple payors, distinguishing it from the Kaiser system. A patient may receive all of their healthcare within Geisinger, or may choose to receive some of their care at Geisinger, while being under

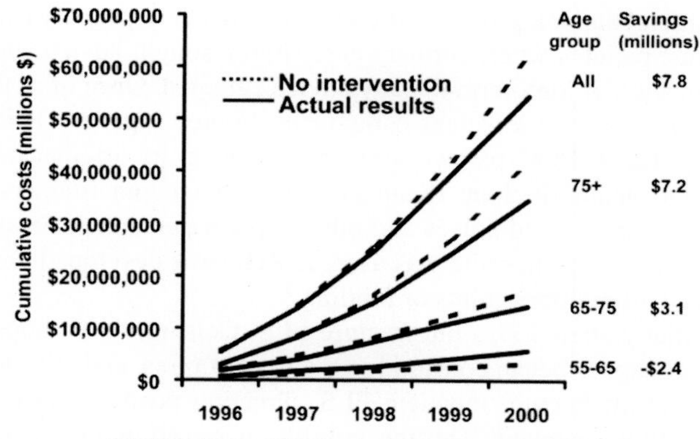

FIGURE 11.5 GHP Projected Healthcare Savings

the care of a non-Geisinger employed primary care physician. Importantly, a pivotal and allied payor is the Geisinger Health Plan, which began as a pilot programme in 1972 and became a formal Health Maintenance Organisation in 1985 and currently has about 600,000 members.[30]

At Geisinger, from 1996 to 2008, an Osteoporosis Disease Management Programme initially focused on community education efforts, peripheral BMD measurement – evolving use of DXA/Vertebral Fracture Assessment (VFA), provider education and disease management guidelines.[31] A System Glucocorticoid-Induced Osteoporosis (GIOP) programme was initiated.[32] A novel Mobile DXA Programme was initiated, providing more convenient and timely DXA services at 22 primary care sites within Geisinger, followed by integration of VFA technology within all system DXA centers.[33,34] An analysis of the impact of this effort showed an estimated $7.8 million dollars in healthcare savings for GHP (Fig. 11.5).[35]

As osteoporosis care evolved, significant efforts were made to communicate with and attempt to engage Geisinger's primary care network through its shared electronic medical record. Despite this documented progress, similar consistencies in clinical care gaps were observed, namely that postfracture patients were not evaluated and/or treated. Geisinger shifted its focus to four groups of patients: (1) at-risk patients not tested; (2) tested patients not accurately risk assessed; (3) high-risk patients not treated and (4) treated patients not adherent.

The first major step was the creation of a uniform outpatient high-risk clinic, termed outpatient HiROC.[7] This outpatient pathway was inaugurated in Geisinger's central campus (Danville) and then integrated in its eastern campus (Wilkes-Barre) and western campus (State College). Additionally, certain team members providing outreach clinic services expanded HiROC access to other primary care sites within Geisinger. Geisinger expected the following groups of outpatients to be referred to outpatient HiROC: fracture patients; patients in need of parenteral therapies, more difficult to administer in primary care clinics; GIOP patients; patients where primary care simply sought advice and patients where the question of long-term therapy advice emerged. Once this clinic pathway was operational, an inpatient component, termed inpatient HiROC, was implemented. The HiROC pathways are illustrated in its original publication and its HiROC teams include rheumatologists, nurse practitioners, clinical nurse specialists, registered nurses and other support personnel. Existing staffing at each regional practice site was used. HiROC was developed without any supplementary programme grants or funding.[7]

While initially started as a hip fracture FLS, Geisinger's Orthopaedic and Hospitalist colleagues began consulting on other fractures, and HiROC quickly evolved into an all-fracture inpatient FLS, directing posthospital care to the Outpatient HiROC pathway. Despite consulting on all fractured inpatients, many HiROC patients decided to follow-up with their own primary care doctors. Given that the records of Geisinger primary care physicians (G-PCP) are available within Epicare,[36] this allowed analysis of postfracture drug initiation in HiROC pathway versus G-PCP. In this analysis, 80.2% of postfracture, drug eligible patients were treated in HiROC compared with only 32.2% in G-PCP care (Table 11.1). This performance data critically show HiROC care outperforms usual G-PCP care, documenting the usefulness and importance of FLS programmes.[7] This data influenced programme design to follow patients longitudinally as opposed to making consultative recommendations and discharging patients to primary care, as some FLS programmes are structured.

Since 2008, HiROC has continued without major changes. A Microsoft Access software database is utilised and HiROC database coordinator helps

TABLE 11.1 Post-Fracture Treatment Comparison (2008–2011) HiROC vs G-PCP

Variable	N=472	HiROC	G-PCP
High-risk treated (%)	337 (71.4%)	308 (80%)	29 (32.2%)
High-risk not treated (%)	122 (25.9%)	70 (18.3%)	52 (57.8%)
Treatment unknown (%)	13 (2.7%)	4 (1.1%)	9 (10%)
Sex			
Female (%)	371 (78.6%)	299 (78.3%)	72 (80%)
Male (%)	101 (21.4%)	83 (21.7%)	18 (20%)
Age			
Mean (years)	77.6 years	76.8 years	81.2 years

G-PCP, geisinger primary care physicians; *HiROC*, high-risk osteoporosis clinic.

facilitate care in the three main HiROC geographic sites. This simple database/registry is different than the Kaiser Permanente information system previously described. This registry is neither programmed to nor capable of identifying real time care gaps as the Kaiser software is. Secondly, to truly measure performance and make programme changes, retrospective chart reviews are necessary. In 2015, Olenginski et al. analysed the first 500 men >50 years old seen in HiROC, from programme startup through January 2015.[37] By this time, 2146 inpatient fracture patients were seen; so, the 500 men made up a significant 23% entire fracture population. Twenty percent of the men died by the 6-month postfracture timepoint, contrasting with the previous reported overall 16% mortality at 6 months of entire HiROC population.[7] DXA scans were done in 82% of these men and HiROC care resulted in 67% of those drug eligible men being treated, once again documenting the effective performance of FLS care.[37]

Recently, both Outpatient HiROC and Inpatient HiROC programme performances were again reappraised. Similar to the initial analysis, a retrospective review documented the following demographics, clinical variables and performance measures: age, sex, fracture site, GHP insurance – yes/no; DXA completed; High-Risk – yes/no; Treatment initiated; 25-OH Vitamin D; deaths – 6 months or later; follow-up visit – yes/no.[38,39]

An Outpatient HiROC analysis from 2014 identified 511 patients seen at all practice sites. Eighty-nine percent of patients were high-risk and drug-eligible. Eighty-nine percent of patients were female with a mean age of 70.0 years. For more than three-quarters of patients, the reason for consult was either T-score or hip/vertebral fracture. Treatment was indicated in 81.3% patients.

Inpatient data on 1279 inpatient consultations from 2013–15 reported mean age to be 77.8 years, and 74% of patients were female. More than two-thirds of patients sustained hip fractures, with the remainder including vertebrae, periprosthetic, subtrochanteric, pelvis, midfemur, distal femur or wrist fractures.

Inpatient 6-month clinical data show 83.6% of this population stratified as high-risk and drug-indicated. Again, 16% of this cohort was deceased at 6-months postfracture. Of the surviving patients, 50% were followed in HiROC, with the rest following with G-PCP or non-Geisinger-PCP (non G-PCP). The superior performance of a dedicated FLS programme of care is documented, with 75.4% of drug-indicated patients being treated in HiROC compared with only 13.8% in G-PCP care. GHP insurance represented 27% of the entire payor mix and, in this group of GHP-insured patients, HiROC care resulted in 75% of drug-indicated patients being treated compared with only 19.7% within G-PCP care. Again, the superior performance of a dedicated FLS programme cannot be overemphasised.[37–39] One surprising clinical care gap found was the identification that 37.8% HiROC patients were subsequently lost to follow-up.[39]

In summary, HiROC care identifies all inpatient fractures by case finding, with use of autoconsult by Hospitalist or Orthopaedic Departments at Geisinger Medical Center, Danville and GWV–Wilkes-Barre. Noteworthy, treatment rates in HiROC of 80%[7] and 75%[39] compared with 32.2% G-PCP[7] and further reduced to 13.8% G-PCP are documented[39] (Table 11.2).

Since Geisinger HiROC programme inception through July 2017, 3196 patients have been seen as inpatient consults and 5411 patients have been seen as outpatient consults.

Despite the maturity of the HiROC programme, care gaps continue to be identified. Patient engagement is often less than ideal. Scheduling and transportation barriers exist posthospitalization. A significant group of patients have needed short-term rehabilitation, and for others, continued loss of independence requires skilled nursing or assisted living arrangements. Facilitation of care is more difficult here, and finding these patients and effectively communicating is challenging. Many patients opt to follow with their primary care physicians, and the use of medications postfracture in this G-PCP group has significantly declined. Patients view available treatments with much more fear and often cannot be convinced to accept treatment to reduce future fracture risk. Although HiROC has never formally studied this, consistent speculations explaining this behaviour include the risk of osteonecrosis of the jaw, atypical femur fractures and the related, heightened media attention. And, even within HiROC pathway, patients are lost to follow-up.[38,39]

Recently, HiROC performance data were presented to Geisinger leadership, including the Chief Medical Officer, GHP leadership and primary care leadership. It was decided to create a workgroup to focus on those patients who are discharged postfracture, refuse HiROC care and follow in G-PCP care. A treatment contract will be shared with the primary care office, Geisinger Care Gap Team and GHP. Geisinger clinical pharmacists will be integrated into strategic primary care sites and offer care to these patients, seeking to attain treatment rates of 50% or more. Additionally, the Geisinger HiROC team is working to better track and prevent lost to follow-up. From an infrastructure standpoint, Geisinger HiROC needs support for development of a new information technology system, one with better functionality, especially to identify gaps in real time and to prospectively track the performance benchmarks already described, eliminating the need for retrospective analyses. Likewise, additional staffing

TABLE 11.2 Post-Fracture Treatment Comparison (2013–2015) HiROC vs G-PCP

	HiROC	G-PCP
Entire population % treated	75.4%	13.8%
GHP-insured % treated	75%	19.7%

G-PCP, Geisinger primary care physicians; *HiROC*, high-risk osteoporosis clinic.

and dedicated time allowing physician FLS champion to more effectively monitor program and performance benchmarks is critical to future success. While a process to help care gaps with GHP-insured patients followed in G-PCP care will soon begin, additional work to narrow care gaps in the other postfractured patients is necessary.

Fracture Liaison Service Demonstration Project

The importance of an effective FLS registry/data base cannot be overemphasised. In December 2013, a collaborative effort began through funding from Merck & Co., Inc. in alliance with CECity, NBHA and NOF.[18,22] Three sites were chosen from 192 sites that applied for this programme support. This 15-month project utilised CECity's Med Concert cloud-based technology platform in hopes of assessing performance measures in FLS care.[18,22]

This project utilised a central database registry, identified and tracked core clinical performance benchmarks in real time, facilitated care from hospital to outpatient clinics and assisted coordinators to facilitate care.

At ASBMR 2016, a pilot study was presented, showing the results of BMD testing, assessment of Vitamin D, prescribing calcium and Vitamin D and prescribing pharmacologic therapy. Participants were patients >50 years with an acute fracture managed in the three participating, independent healthcare systems involved in the demonstration project. Each pilot site serves 450–600 adults hospitalised with low-trauma fractures, and payors, hospitals, patients and physicians were not closely aligned. Using the cloud-based application/registry, the impact of FLS is shown in Fig. 11.6. Substantial gains and improvements in all four metrics are clearly demonstrated.[40]

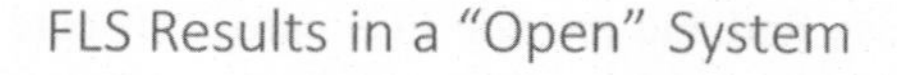

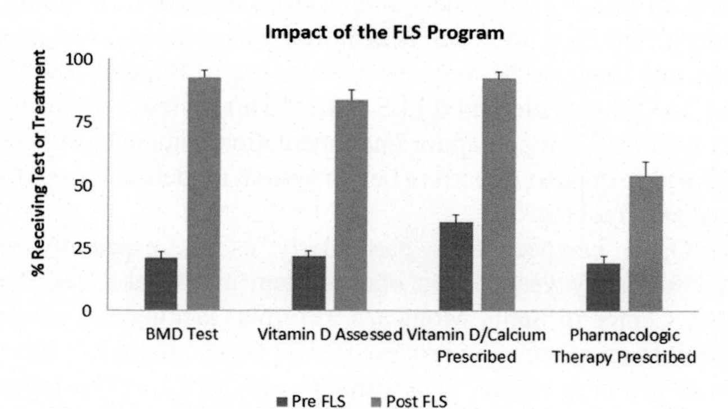

FIGURE 11.6 FLS Demonstration Project Impact

OWN THE BONE

Piloted in 2005 by the AOA, Own the Bone was launched in 2009 and provides a useful clinical tool that helps a site/programme establish an FLS by

1. Helping to identify, evaluate and treat fragility fractures in patients >50 years
2. Help coordinate care
3. Provide a web-based patient registry with 10 prevention measures as well

The components of the registry include demographics, fracture site, fracture history, risk factors, historical medication use, duration of medication use, treatment counselling, treatment initiation, BMD testing, written communication and discharge and follow-up survey.[41,42]

Additionally, the registry allows for the following system reports: subject listings, total #fractured patients, follow-up visit reports, percent follow-up reports, follow-up 'due' reports, follow-up 'dates' report, site fracture reports, letter to patient reports and letter to primary care doctor reports.[41,42]

Similarly, the registry allows the following benchmarking/performance measure reports: Calcium education, Vitamin D education, exercise education, fall prevention, smoking cessation counselling, alcohol education, BMD/DXA testing and treatment initiation.[41,42]

In 2005–06, 14 sites were using this registry. In 2016, Bunta el al documented that 177 sites were registered as Own the Bone registries. Data from the first 125 sites encompassing January 2010 through March 2015 document 23,132 fractured patients in registry. Interventions analysed included education in postfracture setting, development of Own the Bone programme elements, dissemination of information and programme utilisation, implementation and evaluation. After consultation, 53% patients had either DXA testing and/or initiation of pharmacologic treatment.[43]

The Own the Bone initiative can be viewed as an Orthopaedic – centric way of creating a team to case find and provide FLS care. This care can utilise the skills of midlevel providers and/or other members, including primary care providers. This type approach is beautifully summarised and appraised in Miller, Lake, and Emery's 2015 review of Wake Forest Baptist Health's steps in establishing an Orthopaedic-based FLS team.[44] This review very nicely reviews costs/benefits of FLS; a roadmap for implementation; patient workflow; discussion of FLS within context of current health system models and need for system performance and tracking.[44]

Finally, Oates perspective is particularly useful, especially since she had worked for many years in an open system and established the Marian Osteoporosis Center in Santa Monica, California, lacking the advantages of central coordination of care and process flow.[45] Oates poignantly describes the many gaps in her programme, gaps similarly described in Geisinger HiROC programme, in trying to reestablish care after fracturing and being discharged. These include patient education and engagement; documenting the diagnosis of osteoporosis; discharge to skilled nursing home or assisted living facility; difficulty facilitating care, especially for those not discharged to their homes; finding patients in this process; scheduling DXA; losing patients to future follow-up, etc[7,45] Oates reports that 88% of patients surveyed at 1 month posthospitalization were not prescribed medication for their osteoporosis by their primary care physician. She describes communication difficulties, identifying that the lack of physician and medical record continuity is the biggest factor in the care gaps seen in this open FLS setting.[45] The similarities and problems described by this programme correlate with those identified in the Geisinger HiROC programme, an integrated healthcare system FLS, especially since 50% of HiROC patients choose longitudinal HiROC care and follow-up, but the other 50% choose to follow with their primary care physicians.[7,39,45]

Registries/Databases

I have already alluded to the importance of a programme's registry or database. Clearly, again, there is no 'one size fits all' prescriptive advice on registries.

Nonetheless, a registry, should help any FLS programme to

1. Measure work done
2. Coordinate care
3. Capture guideline-driven care
4. Help with follow-up care
5. Assist with tracking functions
6. Monitor FLS benchmarks
7. Ideally detect care gaps

The simplest option is to start with a very simple spreadsheet, perhaps an excel spreadsheet, and determine over time if the success and volume of your programme demands a more sophisticated data capture system. There may be a cost associated with a more sophisticated IT tool.

Many American programmes, especially those Orthopaedic-modelled, have chosen Own the Bone's turnkey registry.[41,42] The main features of this registry and its functionality have already been described.

Likewise, we have discussed the CECity cloud-based registry in discussing the FLS Demonstration Project with NBHA and Medstar Georgetown, UPMC and Creighton Allegiant. There is a cost to this registry and its use, ability to track performance measurements and facilitate care has been reviewed earlier.[18,22]

Kaiser Permanente's custom-programmed registry, the work of Dell and colleagues, has been described earlier and the advantages this registry provides Kaiser[9,10,27] cannot be overemphasised. Many team members (especially Care Managers) and others can use this system, detect gaps in real time and begin to invoke solutions.

The Geisinger HiROC database is a customised Microsoft Access program and has been used since 2008. A database coordinator works behind the scenes to enroll all HiROC patients and to facilitate their care. As HiROC was designed, the idea of a simple, relatively low priced system was paramount, as, akin to many other US programmes, no startup grants or other funding was received. The HiROC team determined which clinical variables would be measured and these same variables used in its initial publication[7] continue to be their key metrics. This database allows the HiROC database coordinator to facilitate care throughout Geisinger HiROC. Its main disadvantages include lack of turnkey capability to monitor and track performance, the inability to detect care gaps in real time and the need for manual, retrospective analyses. The most important Geisinger HiROC metrics continue to include the following: age; sex; fracture site; prior fracture; DXA; 25-OH Vitamin D; Cr; Calcium; Fracture Risk – whether patient is eligible/indicated for treatment; HiROC followup – was it made and is patient adherent/compliant with treatment; PCP followup – is it Geisinger or non-Geisinger primary care; mortality – early or later death. Again, HiROC continues to track these metrics and analyse performance.[7,37–39] Reviewing the available registry options, there is no option easily affordable with ideal functionality. Nonetheless, any programme must enroll, measure, monitor, and track to improve their performance - the registry issue today, represents one of the key and fundamental problems in search of the best solution in US FLS care.

Moving forwards, the FLS movement in the United States could learn from the registries and work of our international colleagues, especially The Royal College of Physicians Fracture Liaison Service Database (FLS-DB).[46,47] Standardised reporting from FLS programmes throughout the world would enable benchmarking against clinical standards and comparisons of performance to be made between services.

QUALITY REPORTING

Mechanisms of reporting are FLS programme specific and are possibly integrated with registry functionality, but not always. Although all FLS programmes are committed to best practice, common threads of metrics and measurements, much duplication exists. Additionally, reporting in the United States conflicts with current 'fee for service' versus pay for performance and quality measures. And, with the changes in US healthcare, it is very difficult for team members of any FLS team to differentiate FLS documentation from those documentation elements that might help with reimbursement and/or quality payments or disincentives.

In a continually changing US healthcare landscape, however, US programmes can consider Physician Quality Reporting System (PQRS); National Committee for Quality Assurance; HEDIS; American Medical Association Physician Consortium for Performance Improvement and Joint Commission. Currently, the only system reporting where potential positive versus negative reimbursement aligns is PQRS.[48–51]

Interestingly, Fojas et al. reported on the effectiveness of two FLS programmes (Nurse Practitioner led vs. physician led) and adherence to Joint Commission measures, including lab testing, DXA and osteoporosis drug treatment in a recent publication.[52] Ideally, if future pay for performance and/or quality metric–based performance and reimbursement could be easily and consistently integrated, then FLS programmes would probably utilise this in their databases and registries.

CONCLUSIONS

It seems clear that a real 'movement' guiding and supporting the FLS pathway of care is on us. Unquestionably, this has begun not only from our learning from our international colleagues but also learning now from each other in the United States. Whether a programme aligns or develops through NBHA, NOF, AOA/Own the Bone, ASBMR, IOF/Capture the Fracture – there is indeed a cooperative and synergistic movement. The continued existence of post-fracture care gaps and the epidemiologic projections as to the magnitude of this problem demand an ongoing spirit of cooperation and collegiality. The NBHA, NOF/IOF, ASBMR, AOA/Own the Bone and others will continue to guide and bolster this movement. Additionally, it seems clear that key stakeholders from corporate America, Pharma, healthcare policymakers and multiple payors will be necessary to aim for a collective solution. Finally, our lawmakers will play a key role, as many of their decisions will both directly and indirectly influence this mission.

If the FLS movement succeeds, it will be clearly evident to see documented expansion and persistence of FLS programmes in every US state, region and territory. And, just as critical, will be the use of documented and key uniform benchmarking and performance metrics – anticipating an era where united and galvanised payors and/or a modified and more effective healthcare system will use this information for future reimbursement decisions and FLS programme support. Importantly, with the ongoing work of secondary fracture prevention efforts, all of us aim to see a documented reduction in the morbidity and mortality and societal impact of osteoporosis-related fractures for the benefit of our patients.

ACKNOWLEDGEMENTS

Undoubtedly, in this writing, I have omitted, forgotten, or had to exclude very important people and players in FLS programmes throughout the United States. For that, I respectfully apologise. Susan Randall, RN, invited me to be a part of the initial ISO "FLS Symposiums" and this encouraged me to work alongside NOF and NBHA over these years. David Lee, Debbie Zeldow, Amy Porter and Elizabeth Thompson have allowed me to continue to work with NBHA and NOF and their respective teams in this effort. Working alongside Dr Rick Dell, Dr Mary Oates, Dr Andrea Singer, Anne Lake, DNP, FNP-BC, ONPC and Kathy Williams EdD has been professionally rewarding and fun. Importantly, whenever I act on behalf of the FLS concept of care, I speak for every member of the Geisinger HiROC team and am respectful of their dedication and loyalty and hard work. Finally, I could not do my work without the love, encouragement, support and inspiration of my wife Jessica and sons Gregory and Lukasz. In concluding, the beauty and humbling nature of this work requires the acknowledgement that we need each other; YES, we need to champion a spirit of collegiality and sharing of best practices - both here and across the seas!

REFERENCES

1. McLellan AR, Galacher SJ, Fraser M, McQuillian C. The fracture liaison service success of a program for the evaluation and management of patients with osteoporotic fracture. *Osteoporos Int* 2003;**14**(12):1028–34.
2. McLellan AR, Wolowacz SE, Zimovetz EA, Bead SM, Lock S, McCrink L, Adekunle F, Roberts D. Fracture liaison services for the evaluation and management of patients with osteoporotic fracture: a cost-effectiveness evaluation based on data collected over 8 years of service provision. *Osteoporos Int* 2011;**22**:2083–98.
3. Bogoch ER, Elliot-Gibson V, Beaton DE, Jamal SA, Josse RG, Murray TM. Effective initiation of osteoporosis diagnosis and treatment for patients with a fragility fracture in an orthopaedic environment. *J Bone Joint Surg Am* 2006;**88**(1):25–34.
4. Sale JEM, Beaton D, Posen J, Elliot-Gibson V, Bogoch E. Systematic review on interventions to improve osteoporosis investigation and treatment in fragility fracture patients. *Osteoporos Int* 2011;**22**:2067–82.
5. Mitchell PJ. Best practices in secondary fracture prevention: fracture liaison services. *Curr Osteoporos Rep* 2013;**11**:52–60.
6. Akeeson K, Marsh D, Mitchell PJ, McLellan AR, Stenmark J, Pierroz DD, Kyer C, Cooper C, IOF Fracture Working Group. capture the fracture: a best practice framework and global campaign to break the fragility fracture cycle. *Osteoporos Int* 2013;**24**:2135–52.
7. Olenginski TP, Maloney-Saxon G, Matzko CM, Mackiewicz K, Kirchner HL, Bengier A, Newman ED. High-risk osteoporosis clinic (HiROC): improving osteoporosis and postfracture care with an organized, programmatic approach. *Osteoporos Int* 2015;**26**(2):801–10.
8. https://www.ncbi.nlm.nih.gov/pubmed/21107534.
9. Dell R. Fracture prevention in Kaiser Permanente Southern California. *Osteoporos Int* 2011;**22**(Suppl. 3):S457–60.
10. Dell R, Greene D, Schelkun SR, Williams K. Osteoporosis disease management: the role of the orthopaedic surgeon. *J Bone Joint Surg Am* 2008;**90**(Suppl. 4):188–94.
11. Briot K. Fracture liaison services. *Curr Opin Rheumatol* 2017;**29**:416–21.
12. Giangregorio L, Papaioannou A, Cranney A, Zytaruk N, Adachi JD. Fragility fractures and the osteoporosis care gap: an international phenomenon. *Semin Arthritis Rheum* 2006;**35**:293–305.
13. Port L, Center J, Briffa NK, Nguyen T, Cumming R, Eisman J. Osteoporotic fracture: missed opportunity for intervention. *Osteoporos Int* 2003;**14**:780–4.
14. Edwards BJ, Bunta AD, Simonelli C, Bolander M, Fitzpatrick LA. Prior fractures are common in patients with subsequent hip fractures. *Clin Orthop Relat Res* 2007;**461**:226–30.
15. Office of the Surgeon General (US). *Bone health and osteoporosis: a report of the Surgeon General* Rockville (MD). 2004.
16. Khosla S, Cauley JA, Compston J, Kiel DP, Rosen C, Saag KG, Shane E. Addressing the crisis in the treatment of osteoporosis: a path forward. *JBMR* 2017;**32**(3):424–30.
17. Tosi LL, Gliklich R, Kannan K, Koval KJ. The American Orthopaedic Association's "own the bone" initiative to prevent secondary fractures. *J Bone Joint Surg Am* 2008;**90**(1):163–73.

18. http:www.nbha.org.
19. http://www.2million2many.org.
20. www.nof.org.
21. www.capturethefracture.org.
22. Holzmueller CG, Karp S, Zeldow D, Lee DB, Thompson DA. Development of a cloud-based application for the fracture liaison service model of care. *Osteoporos Int* 2016;**27**(2): 683–90.
23. https://en.wikipedia.org/wiki/Kaiser_Permanente.
24. http://www.achp.org/wp-content/uploads/Kaiser-Permanente-Innovation-Profile-ACHP.pdf.
25. https://healthy.kaiserpermanente.org/.
26. https://healthy.kaiserpermanente.org/static/health/pdfs/quality_and_safety/sca/sca_quality_HEDIS.pdf.
27. Greene D, Dell R. Outcomes of an osteoporosis disease - management program managed by nurse practitioners. *J Am Acad Nurse Pract* 2010;**22**(6):326–9.
28. https://en.wikipedia.org/wiki/Geisinger_Health_System.
29. www.geisinger.org.
30. www.geisinger.org/health-plan.
31. Newman ED. Perspectives on pre-fracture intervention strategies: the Geisinger Health System osteoporosis program. *Osteoporos Int* 2011;**22**(Suppl. 3):S451–5.
32. Newman ED, Matzko CM, Olenginski TP, Perruquet JL, Harrington TM, Maloney-Saxon G, Culp T, Wood GC. Glucocorticoid-induced osteoporosis program (GIOP): a novel, comprehensive, and highly successful program with improved outcomes at 1 year. *Osteoporos Int* 2006;**17**:1428–34.
33. Newman ED, Olenginski TP, Perruquet JL, Hummel J, Indeck C, Wood GC. Using mobile DXA to improve access to osteoporosis care. *J Clin Densitom* 2004;**7**:71–5.
34. Olenginski TP, Newman ED, Hummel JL, Hummer M. Development and evaluation of a vertebral fracture assessment program using IVA and its integration with mobile DXA. *J Clin Densitom* 2006;**9**(1):72–7.
35. Newman ED, Ayoub WT, Starkey RH, Diehl JM, Wood GC. Osteoporosis disease management in a rural health population: hip fracture reduction and reduced costs in postmenopausal women after 5 years. *Osteoporos Int* 2003;**14**:146–51.
36. www.epic.com/software.
37. Olenginski TP, et al. *Analysis of 500 men seen in HiROC*. Presented at ISO Washington, DC 2015. 2015.
38. Dunn PD, Webb D, Olenginski T. *Outpatient HiROC program analysis 2014*. Presented at Clinical Osteoporosis 2017, Orlando, FL. 2017.
39. Dunn P, Webb D, Olenginski TP. Geisinger high-risk osteoporosis clinic (HiROC): 2013-2015 FLS performance analysis. *Osteoporos Int* 2018;**29**:451–7.
40. Greenspan S, Singer A, Lee D, et al. *FLS model in an open system*. Presented at ASBMR 2016, Atlanta, GA. 2016.
41. www.ownthebone.org.
42. www.aoassn.org/aoaimis/OTB/About/What_Is_Own_the_Bone.aspx.
43. Bunta AD, et al. Own the bone, a system-based intervention, improves osteoporosis care after fragility fractures. *J Bone Joint Surg Am* 2016;**98**(24):e109.
44. Miller AN, Lake AF, Emery CL. Establishing a fracture liaison service:an orthopaedic approach. *J Bone Joint Surg Am* 2015;**97**:675–81.
45. Oates MK. Invited commentary: fracture follow-up program in an open health care system. *Curr Osteoporos Rep* 2013;**11**:369–76.

46. Javaid MK, Boulton C, Gallagher C, Judge A, Vasilakis N. Fracture Liaison Service Database (FLS-DB) annual report: Leading FLS improvement: secondary fracture prevention in the NHS. In: Physicians RCo, editor. *Healthcare Quality Improvement Partnership*. 2017. London. https://www.fffap.org.uk/fls/flsweb.nsf.
47. Javaid MK, Kyer C, Mitchell PJ, et al. Effective secondary fracture prevention: implementation of a global benchmarking of clinical quality using the IOF capture the fracture(R) best practice framework tool. *Osteoporos Int* 2015;**26**:2573–8.
48. https://qpp.cms.gov/.
49. http://www.ncqa.org/hedis-quality-measurement.
50. https://ecqi.healthit.gov/measure-stewards/american-medical-association-convened-physician-consortium-performance-improvementr.
51. https://www.jointcommission.org/improving_and_measuring_osteoporosis_management/.
52. Fojas MC, Southerland LT, Phieffer LS, Stephens JA, Srivastava T, Ing SW. Compliance to the Joint Commission proposed core measure set on osteoporosis-associated fracture: review of different secondary fracture prevention programs in an open medical system from 2010 to 2015. *Arch Osteoporos* 2017;**12**:16.

Chapter 12

National and International Programs

Paul J. Mitchell[1,2]
[1]School of Medicine, Sydney Campus, University of Notre Dame Australia, Sydney, NSW, Australia; [2]Osteoporosis New Zealand, Wellington, New Zealand

AUSTRALIA

FLS/SFPP Implementation Initiatives

In 2016, the Australian and New Zealand Bone and Mineral Society launched a secondary fracture prevention program (SFPP) initiative.[1] This initiative provided clinicians and administrators with a suite of resources to support development of new programs or refinement of those already established. This initiative was based on the approach taken by Osteoporosis Canada described below.[2]

In 2015, representatives of 22 organisations gathered in Sydney to attend the inaugural National Forum on Secondary Fracture Prevention.[3] The Forum served as a call to action, with the intention of engaging all stakeholders to advocate for a national approach to secondary fracture prevention. In 2016, the Stop Osteoporosis Secondary (SOS) Fracture Alliance was formed to advance the objectives identified in the Forum.[4] As of September 2017, the SOS Fracture Alliance unites 31 medical, allied health, patient advocacy, carer and other organisations under its umbrella. The more than 2.9 million members have one common goal—to 'make the first break the last' by improving the care of patients presenting with an osteoporotic fracture. This is the first time in Australia an alliance of organisations has formed to address this public health issue across the nation.

Clinical Standards

In 2014, the New South Wales Agency for Clinical Innovation published *Minimum Standards for the Management of Hip Fracture in the Older Person.*[5] Standard 6 relates to secondary fracture prevention. In 2016, the Australian Commission on Safety and Quality in Health Care, in collaboration with the Health Quality and Safety Commission New Zealand published the *Hip Fracture Care Clinical Care Standard.*[6] The standard includes seven quality

Secondary Fracture Prevention. https://doi.org/10.1016/B978-0-12-813136-7.00012-0

statements. Quality statement 6 relates to provision of a falls and bone health assessment prior to discharge of the hip fracture patient from hospital. Current performance against these standards is documented in the chapter by Close on Oceania.

Fracture Registries

The Australian and New Zealand Hip Fracture Registry (ANZHFR) was established in 2015.[7] As of June 2017, the number of sites in Australia with ethics and governance approvals to participate in this registry was 46.[8] Of these, 80% were utilising the registry by entering date on hip fracture patients and 20% were preparing to implement data collection. The Australian arm of the registry held more than 10,000 records at the end of June 2017. The 2017 Annual Report of the ANZHFR included the following findings relating to secondary fracture prevention[9]:

- A third of hospitals in Australia and New Zealand reported having an Fracture Liaison Service (FLS)/SFPP. Among the 97 hospitals from Australia who participated in the facilities level audit, 26 (27%) reported having an FLS/SFPP (Personal communication: E. Armstrong). Of note, only hospitals performing hip fracture surgery could participate in the survey.
- 16% of patients left hospital on a bisphosphonate, denosumab, or teriparatide compared to 8% on admission.
- 78% of patients underwent a fall risk assessment during their in-patient stay.

More information on the ANZHFR is provided in Chapter 4.

CANADA

FLS/SFPP Implementation Initiatives

In 2013, Osteoporosis Canada launched their campaign *Make the FIRST break the LAST with Fracture Liaison Services.*[2] This initiative was based around a summary document and call to action, supported by a comprehensive suite of appendices. In addition to this national initiative, the Ontario Osteoporosis Strategy (OOS) was launched in 2005 and comprised five key components which include postfracture care.[10] The OOS is an initiative of the Ontario Chronic Disease Prevention and Management Strategy and is funded by the Ontario Ministry of Health and Long-term Care. The initial investment was CN$5 million annually. The postfracture component of the strategy includes a province-wide Fracture Clinic Screening Program delivered by screening coordinators working in high and medium volume fracture clinics. This program was developed and operated by Osteoporosis Canada in partnership with the Ontario Orthopaedic Association and the Ontario College of Family Physicians. The impact of this program is described in Chapter 5.

Clinical Standards

In 2014, Osteoporosis Canada published *Quality Standards for Fracture Liaison Services in Canada* which described seven key standards for an FLS to deliver.[11] In 2016, Osteoporosis Canada published *Essential Elements of Fracture Liaison Services* which described eight key characteristics of a high-performing FLS.[12] Also in 2016, Osteoporosis Canada launched an FLS Registry,[13] an online map profiling FLS programmes across Canada, which meets Osteoporosis Canada's Essential Elements. As of August 2017, 41 FLSs feature on the registry.

Fracture Registries

Although a national hip fracture registry has not been established in Canada, a registry is under development in British Columbia. An analysis of care delivered to 8000 hip fracture patients in British Columbia was presented at the International Forum on Quality and Safety in Healthcare in Kuala Lumpur, Malaysia in August 2017.[14]

NEW ZEALAND

FLS/SFPP Implementation Initiatives

In 2012, Osteoporosis New Zealand published *BoneCare 2020: A systematic approach to hip fracture care and prevention for New Zealand.*[15] This strategy identified four key objectives:

1. Improve outcomes and quality of care after hip fractures by delivering professional standards of care monitored by a recently introduced New Zealand National Hip Fracture Registry.
2. Respond to the first fracture to prevent the second through universal access to Fracture Liaison Services in every District Health Board in New Zealand.
3. General Practitioners to stratify fracture risk within their practice population using fracture risk assessment tools supported by local access to axial bone densitometry.
4. Consistent delivery of public health messages on preserving physical activity, healthy lifestyles and reducing environmental hazards.

The strategy called on all stakeholder organisations to join a National Fragility Fracture Alliance to support implementation. Subsequently, Osteoporosis New Zealand, the Ministry of Health, Accident Compensation Corporation, Health Quality and Safety Commission New Zealand, the ANZHFR and all relevant learned societies collaborated. Key initiatives included:

- The Ministry of Health, Osteoporosis New Zealand, the four regional District Health Board Alliances (and their clinical and administrative staff) delivered FLS Forums to share best practice and experience from elsewhere during

Q4-2013 and Q1-2014.[16] In 2013, the Ministry of Health District Annual Planning guidance for 2014–15 stated that the District Health Boards should have fully operational FLS, and that implementation would be measured quarterly.[17]

- Osteoporosis New Zealand developed a suite of resources to support District Health Boards to implement FLS, including[18]:
  - FLS Resource pack: A guide to relevant national policy and strategies, the rationale for secondary fracture prevention, the management gap in New Zealand and description of operational aspects of FLS.
  - FLS Status summary.
  - FLS Business plan template.
  - FLS Step-by-step guide.
  - Generic Fracture Liaison Nurse job description.
- In July 2016, the Accident Compensation Corporation announced an investment of NZ$30.5 million over 4 years to reduce falls and fractures among older New Zealanders.[19] The investment funded access to:
  - In-home and community-based strength and balance programmes.
  - FLS, to identify and treat those at risk of osteoporosis and further fractures.
  - Assessment and management of visual acuity and environmental hazards in the home.
  - Medication review for people taking multiple medicines.
  - Vitamin D prescribing in Aged Residential Care.
  - Integrated services across primary and secondary care (including supported hospital discharge) to provide seamless pathways in the falls and fracture system.
- In December 2016, the Ministry of Health published the *Healthy Ageing Strategy* which advocated implementation of evidence-based models of care to improve the quality of care for those admitted for falls and fractures, including hip fractures.[20]
- In May 2017, the *Live Stronger for Longer* initiative was launched by Accident Compensation Corporation, Ministry of Health, Health Quality and Safety Commission and the New Zealand Government, in collaboration with all relevant consumer groups and District Health Boards.[21] This consumer-focused initiative creates a common identity for the various components of the national approach to falls and fracture reduction.
- In June 2017, the first Falls and Fractures Outcomes Framework quarterly report was published.[22] Fig. 12.1, which appeared in the second quarterly report,[23] describes five domains which are populated with a range of measures pertaining to falls and fracture care.

Clinical Standards

In 2016, as described above for Australia, the Australian Commission on Safety and Quality in Health Care, in collaboration with the Health Quality and Safety

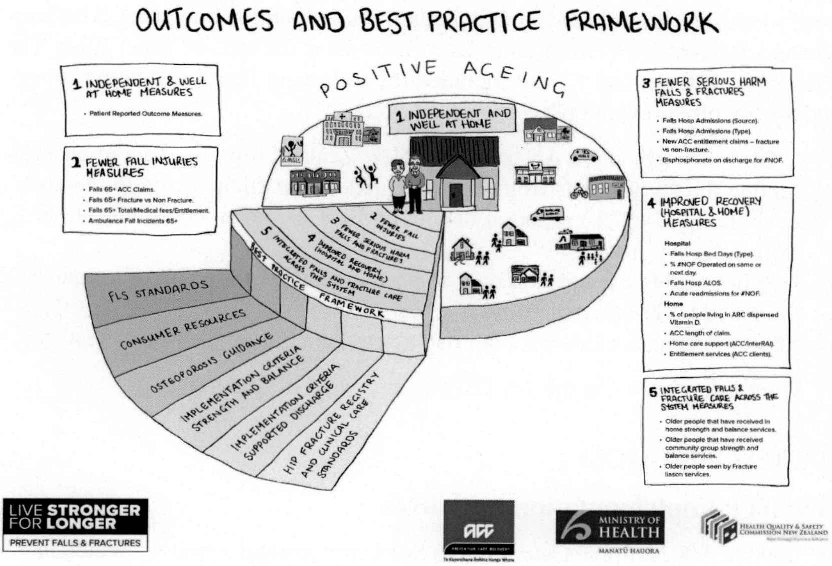

FIGURE 12.1 Falls and Fractures Outcomes Framework for New Zealand.[23]

Commission New Zealand published the *Hip Fracture Care Clinical Care Standard*.[6] The standard includes seven quality statements. Quality statement 6 relates to provision of a falls and bone health assessment prior to discharge of the hip fracture patient from hospital.

Also in 2016, Osteoporosis New Zealand published *Clinical Standards for Fracture Liaison Services in New Zealand*.[24] The six key standards were based on the approach taken by the UK National Osteoporosis Society described below,[25] and they are aligned to the International Osteoporosis Foundation's 'Capture the Fracture®' Best Practice Framework Standards.[26–28] A broad consultation exercise was undertaken, inviting critique from all relevant learned societies and organisations in New Zealand, the International Osteoporosis Foundation and the global Fragility Fracture Network.[29] The clinical standards were subsequently endorsed by 15 organisations, including the Accident Compensation Corporation, Health Quality and Safety Commission, International Osteoporosis Foundation and Fragility Fracture Network.

Fracture Registries

As described for Australia above, the ANZHFR was established in 2015.[7] As of June 2017, the number of sites in New Zealand with ethics and locality approval to participate in the ANZHFR was 17.[8] Of the 20 District Health Boards in New Zealand, 60% were utilising the Registry by entering data on hip fracture patients, 20% were preparing to implement data collection and 20%

were interested in progressing approvals to collect and submit data. The New Zealand Registry held more than 2200 records at the end of June 2017. The 2017 ANZHFR Annual Report included the following findings relating to secondary fracture prevention[9]:

- A third of hospitals in Australia and New Zealand reported having an FLS. Among the 23 hospitals from New Zealand who participated in the facilities level audit, 13 (57%) reported having an FLS (Personal communication: E. Armstrong).
- 31% of patients left hospital on a bisphosphonate, denosumab or teriparatide compared to 8% on admission.
- 90% of patients underwent a fall risk assessment during their in-patient stay.

More information on the ANZHFR is provided in Chapter 4.

UNITED KINGDOM

FLS/SFPP Implementation Initiatives

In 2009, the UK National Osteoporosis Society published a strategy which identified five priority areas, the first of which being implementation of an FLS for every hospital.[30] In the same year, the Department of Health for England published *Falls and fractures: Effective interventions in health and social care*.[31] This policy document identified four key objectives:

1. Improve outcomes and improve efficiency of care after hip fractures—by following the 6 'Blue Book' standards (described below).[32]
2. Respond to the first fracture, prevent the second—through Fracture Liaison Services in acute and primary care.
3. Early intervention to restore independence—through falls care pathway linking acute and urgent care services to secondary falls prevention.
4. Prevent frailty, preserve bone health, reduce accidents—through preserving physical activity, healthy lifestyles and reducing environmental hazards.

The Department of Health subsequently published *Fracture prevention services: an economic evaluation* to support the case for widespread implementation of FLS.[33] The National Osteoporosis Society also developed a suite of online resources, including an FLS Benefits Calculator, and established a Service Development Team which is working with hospitals throughout the United Kingdom to improve access to the FLS.[34]

Clinical Standards

In 2007, the British Orthopaedic Association and the British Geriatrics Society jointly led development of consensus guidelines for hip fracture care and prevention.[32] The 'Blue Book' titled *The care of patients with fragility fracture* proposed six standards. Standards 5 and 6 related to osteoporosis assessment and falls assessment, respectively. In 2012, the National Institute for Heath and Care Excellence (NICE) published Quality Standard 16 relating to hip fracture care, which included 12 quality statements.[35] Quality statements 11 and 12 related to falls risk assessment and osteoporosis risk assessment, respectively.

In 2015, the National Osteoporosis Society published *Effective Secondary Prevention of Fragility Fractures: Clinical Standards for Fracture Liaison Services*.[25] The clinical standards were based on the so-called '5IQ' approach:

- Identification: Case-finding fracture patients.
- Investigation: Fracture and falls risk assessment.
- Information: Educating patients on their fracture and falls risk.
- Intervention: Guidelines-based use of drug treatments and nonpharmacological options.
- Integration: Provision of long-term management plans to general practitioners and other health-care professionals.
- Quality: Ongoing audit of the FLS, continuing professional development and peer review.

The clinical standards were subject to a public consultation and subsequently endorsed by nine organisations: In 2016, the University Hospital Birmingham FLS team benchmarked care delivered by their service against the National Osteoporosis Society clinical standards.[36] The authors concluded that the standards could be delivered in the context of a busy teaching hospital.

In 2017, NICE published Quality Standard 149 which relates to osteoporosis.[37] Quality Statement 1 states that 'Adults who have had a fragility fracture or use systemic glucocorticoids or have a history of falls have an assessment of their fracture risk'. Quality Statement 2 states 'Adults at high risk of fragility fracture are offered drug treatment to reduce fracture risk'.

Fracture Registries

The UK National Hip Fracture Database was established in 2007 and is currently the largest continuous audit of acute hip fracture care and secondary fracture prevention after hip fracture in the world.[38] The 2016 Annual Report by the National Hip Fracture Database included the following findings relating to secondary fracture prevention[39]:

- 97.2% of patients had been assessed for the need for bone protection medication.
- 79.3% of patients had been started on bone protection medication prior to discharge, or referred for bone density testing or a bone clinic appointment, or were already on appropriate medication.
- 97.0% of patient had a multidisciplinary assessment to identify and address future risk of falling.

In 2015, the FLS Database was established by the Royal College of Physicians,[40] and its first patient level audit was published in 2017.[41] Key findings from this audit included:

- Information on 18,356 patients was entered by 38 participating FLSs.
- National coverage of secondary fracture prevention by the FLS was low.
- The variability in quality between the existing FLS highlighted the need for continuous national audit of secondary fracture prevention.

More information on the registry initiatives is provided in Chapter 10.

UNITED STATES OF AMERICA

Secondary fracture prevention efforts in the United States are being led nationally by the National Bone Health Alliance,[42] a public–private partnership launched in 2010 that brings together the expertise and resources of its member organisations to collectively: promote bone health and prevent disease; improve diagnosis and treatment of bone disease and enhance bone research, surveillance and evaluation. In 2013, the National Bone Health Alliance established the Fracture Prevention CENTRAL website which provides a comprehensive suite of resources to support implementation of FLSs.[43] An associated webinar series shares experience from established high-performing FLSs with health professionals and administrators who are developing services.

Clinical Standards

Clinical standards relating to secondary fracture prevention are yet to be developed in the United States.

Fracture Registries

In 2014, the National Osteoporosis Foundation and the National Bone Health Alliance launched a Qualified Clinical Data Quality Improvement Registry.[44] This allows eligible professionals to submit their quality measures to Centers for Medicare and Medicaid Services to meet their Physician Quality Reporting System quality reporting requirements.

More information on the various US initiatives is provided in the chapter by Olenginski on the USA.

INTERNATIONAL INITIATIVES

International Osteoporosis Foundation Capture the Fracture Program

In 2012, the International Osteoporosis Foundation launched the Capture the Fracture Program with publication of the 2012 World Osteoporosis Day thematic report.[45] Since 2012, Capture the Fracture has developed into one of the Foundation's leading initiatives. The key components of Capture the Fracture are:

- Website: The Capture the Fracture website provides a comprehensive suite of resources to support health-care professionals and administrators to establish a new FLS or improve an existing FLS.
- Webinars: An ongoing series of webinars provide an opportunity to learn from experts across the globe who have established high-performing FLSs and contributed to development of guidelines and policy on secondary fracture prevention. As of August 2017, webinars have been conducted in Chinese, Dutch, English, French, Italian, Japanese, Polish, Portuguese and Spanish.
- Best Practice Framework: The Best Practice Framework, which is currently available in nine major languages, sets an international benchmark for the FLS by defining essential and aspirational elements of service delivery. This framework serves as the measurement tool for the International Osteoporosis Foundation to award Capture the Fracture Best Practice Recognition status. The 13 globally endorsed standards of the framework were published in 2013 and are as follows[27]:
    - Patient Identification Standard
    - Patient Evaluation Standard
    - Post-fracture Assessment Timing Standard
    - Vertebral Fracture Standard
    - Assessment Guidelines Standard
    - Secondary Causes of Osteoporosis Standard
    - Falls Prevention Services Standard
    - Multifaceted health and lifestyle risk-factor Assessment Standard
    - Medication Initiation Standard
    - Medication Review Standard
    - Communication Strategy Standard
    - Long-term Management Standard
    - Database Standard

The Best Practice Framework tool has been tested in a range of health settings across the world.[28] As of August 2017, 233 FLS feature on the Map of Best Practice.

Fragility Fracture Network

The Fragility Fracture Network has kindly granted permission for the following content to be reproduced from the network's Strategic Plan[46] and website.[29]

The Fragility Fracture Network is a global organisation, which was founded to create a multidisciplinary network of experts for improving treatment and secondary prevention of fragility fractures. The Network's mission and vision underline this and are the foundation for all decisions taken.

The Fragility Fracture Network believes that useful policy change can only happen at a national level, and multidisciplinary national coalitions are the most effective way to achieve this. Hence the Network acts as a global template for creating national alliances in as many countries as possible. The strategic goals of the Fragility Fracture Network are as follows:

- To create a global network of national alliances of fragility fracture activists.
- To spread globally the best multidisciplinary practice and systems of care for managing fragility fractures.
- To ensure that every fragility fracture becomes an opportunity for systematically preventing further fracture.
- To promote research aimed at improving the quality of fragility fracture care.
- To generate political priority for fragility fracture care in all countries.

For the next 5 years, the strategic focus of the Fragility Fracture Network is to facilitate national (or regional) multidisciplinary alliances which lead to:

- Consensus guidelines,
- Quality standards and
- Systematic performance measurements

for the care of older people with fragility fracture. The metric of Fragility Fracture Network's success will be the number of nations in which these goals are achieved.

REFERENCES

1. Australian and New Zealand Bone and Mineral Society. *Secondary fracture prevention program initiative*. 2016. http://www.fragilityfracture.org.au/.
2. Osteoporosis Canada. *Make the FIRST break the LAST with fracture liaison services*. 2013.
3. Australian and New Zealand Bone and Mineral Society. *First national forum on secondary fracture prevention: meeting report* Sydney. 2015.
4. SOS Fracture Alliance. *SOS Fracture Alliance: making the first break the last*. 2017. https://www.sosfracturealliance.org.au/.
5. New South Wales Agency for Clinical Innovation. *Minimum standards for the management of hip fracture in the older person*. Chatswood, NSW. 2014.
6. Australian Commission on Safety and Quality in Health Care, Health Quality & Safety Commission New Zealand. *Hip fracture care clinical care standard*. Sydney. 2016.
7. Australian and New Zealand Hip Fracture Registry. *Australian and New Zealand Hip Fracture Registry website*. 2017. http://www.anzhfr.org/.
8. Australian and New Zealand Hip Fracture Registry. *ANZ Hip Fracture Registry newsletter June 2017*. Sydney. 2017.
9. Australian and New Zealand Hip Fracture Registry. *Annual report 2017* Sydney. 2017.
10. Jaglal SB, Hawker G, Cameron C, et al. The Ontario Osteoporosis Strategy: implementation of a population-based osteoporosis action plan in Canada. *Osteoporos Int* 2010;**21**(6):903–8.
11. Osteoporosis Canada. *Quality standards for fracture liaison services in Canada*. Toronto: Osteoporosis Canada; 2014.
12. Osteoporosis Canada. *Essential elements of fracture liaison services (FLS)*. Toronto. 2016.

13. Osteoporosis Canada. *FLS registry map*. 2017. https://www.osteoporosis.ca/fls/fls-tools-and-resources/fls-registry-map/.
14. Sobolev B. F3: hip fracture care redesign in British Columbia: 3-year follow-up. In: *International forum on quality and safety in healthcare; 24–26 August 2017, Kuala Lumpur Malaysia*. 2017.
15. Osteoporosis New Zealand. *Bone care 2020: a systematic approach to hip fracture care and prevention for New Zealand*. Wellington. 2012.
16. Mitchell PJ, Cornish J, Milsom S, et al. BoneCare 2020: a systematic approach to hip fracture care and prevention for New Zealand. In: *3rd fragility fracture network congress 2014; 4–6 September 2014, Madrid, Spain*. 2014.
17. Ministry of Health. *Annual plan guidance: working draft 2014/15 toolkit annual plan with statement of intent*. Wellington. 2013.
18. Osteoporosis New Zealand. *Fracture liaison services*. 2014. https://osteoporosis.org.nz/resources/health-professionals/fracture-liaison-services/.
19. New Zealand Government. *ACC invests $30m to reduce falls and fractures for older New Zealanders*. 2016. https://www.beehive.govt.nz/release/acc-invests-30m-reduce-falls-and-fractures-older-new-zealanders.
20. Associate Minister of Health. In: Ministry of Health, editor. *Healthy ageing strategy*. 2016. Wellington.
21. Accident Compensation Corporation, Ministry of Health, Health Quality & Safety Commission New Zealand, New Zealand Government. *Live stronger for longer: prevent falls and fractures*. 2017. http://livestronger.org.nz/.
22. Accident Compensation Corporation, Ministry of Health, Health Quality & Safety Commission New Zealand. *Falls and fractures outcomes framework quarterly report - April 2017*. Wellington: Accident Compensation Corporation; 2017.
23. Accident Compensation Corporation, Ministry of Health, Health Quality & Safety Commission New Zealand. *Falls and fractures outcomes framework quarterly report - July 2017*. Wellington: Accident Compensation Corporation; 2017.
24. Osteoporosis New Zealand. *Clinical standards for fracture liaison services in New Zealand*. Wellington: Osteoporosis New Zealand; 2016.
25. Gittoes N, McLellan AR, Cooper A, et al. *Effective secondary prevention of fragility fractures: clinical standards for fracture Liaison services*. Camerton: National Osteoporosis Society; 2015.
26. International Osteoporosis Foundation. *Capture the fracture® programme website*. 2017. http://www.capture-the-fracture.org/.
27. Akesson K, Marsh D, Mitchell PJ, et al. Capture the Fracture: a Best Practice Framework and global campaign to break the fragility fracture cycle. *Osteoporos Int* 2013;**24**(8):2135–52.
28. Javaid MK, Kyer C, Mitchell PJ, et al. Effective secondary fracture prevention: implementation of a global benchmarking of clinical quality using the IOF Capture the Fracture(R) Best Practice Framework tool. *Osteoporos Int* 2015;**26**(11):2573–8.
29. Fragility Fracture Network. *Fragility Fracture Network website*. 2017. http://fragilityfracturenetwork.org/.
30. National Osteoporosis Society. *Protecting fragile bones: a strategy to reduce the impact of osteoporosis and fragility fractures in England/Scotland/Wales/Northern Ireland. May–Jun 2009*. 2009.
31. Department of Health. In: Department of Health, editor. *Falls and fractures: effective interventions in health and social care*. 2009.
32. British Orthopaedic Association, British Geriatrics Society. *The care of patients with fragility fracture*. 2007.

33. Department of Health. In: *Fracture prevention services: an economic evaluation*. 2009.
34. National Osteoporosis Society. *Fracture liaison services*. 2017. https://nos.org.uk/for-health-professionals/service-development/fracture-liaison-services/.
35. National Institute for Health and Clinical Excellence. *Quality standard for hip fracture care. NICE Quality Standard 16*. London. 2012.
36. Shipman KE, Stammers J, Doyle A, Gittoes N. Delivering a quality-assured fracture liaison service in a UK teaching hospital-is it achievable? *Osteoporos Int* 2016;**27**(10):3049–56.
37. National Institute for Health and Care Excellence. *Quality standard 149*. London. Osteoporosis; 2017.
38. Royal College of Physicians. *The national hip fracture database*. 2017. http://www.nhfd.co.uk/.
39. Royal College of Physicians. *National hip fracture database (NHFD) annual report 2016*. London: RCP; 2016.
40. Royal College of Physicians. *Fracture liaison service database (FLS-DB)*. 2017. https://www.rcplondon.ac.uk/projects/fracture-liaison-service-database-fls-db.
41. Royal College of Physicians. *FLS-DB clinical audit: identifying high-quality care in the NHS for secondary fracture prevention*. London. 2017.
42. Lee DB, Lowden MR, Patmintra V, Stevenson K. National Bone Health Alliance: an innovative public-private partnership improving America's bone health. *Curr Osteoporos Rep* 2013;**11**(4):348–53.
43. National Bone Health Alliance. *Fracture prevention CENTRAL website*. 2017. http://www.nbha.org/fpc.
44. National Osteoporosis Foundation, National Bone Health Alliance, CE City. *The NOF and NBHA qualified clinical data quality improvement registry (QCDR)*. 2016. https://www.medconcert.com/content/medconcert/FractureQIR/.
45. Akesson K, Mitchell PJ. *Capture the fracture: a global campaign to break the fragility fracture cycle*. Nyon: International Osteoporosis Foundation; 2012.
46. Fragility Fracture Network. *Strategic plan 2017–2021*. Berlin. 2017.

Appendix

Summary of Useful Resources

Country/ Region	Organisation	Resource	Links
Australia	Australian and New Zealand Bone and Mineral Society	Secondary Fracture Prevention Program Initiative: A suite of resources including analysis of Australian FLS programs, a business plan template and a step-by-step guide to setting up programmes.	http://www.fragility fracture.org.au/
	Australian National SOS Fracture Alliance	Australia's largest Alliance for the promotion of secondary fracture prevention. Strategic plan and many links.	http://www.sosfracture alliance.org.au
Canada	Osteoporosis Canada	FLS Hub: A comprehensive web-based resource including the Canadian FLS Registry, an FLS Toolkit, Key Indicators, Quality Standards, a free consultation service and a suite of webinars.	http://fls.osteoporosis.ca/
New Zealand	Osteoporosis New Zealand	FLS web pages: A suite of resources including an FLS Resource Pack, a business plan template, generic Fracture Liaison Nurse job descriptions and clinical standards for FLS.	https://osteoporosis.org.nz/resources/health-professionals/fracture-liaison-services/ https://osteoporosis.org.nz/resources/health-professionals/clinical-standards-for-fls/

Country/ Region	Organisation	Resource	Links
United Kingdom	National Osteoporosis Society	FLS web pages: A suite of resources including an FLS Implementation Toolkit (with a business case benefits calculator, outcomes and performance indicators), clinical standards and case studies of exemplar services.	https://nos.org.uk/for-health-professionals/service-development/fracture-liaison-services/
United States of America	National Bone Health Alliance	Fracture Prevention CENTRAL website: A comprehensive web-based resource which includes business case templates, a return on investment calculator, a webinar series and the opportunity to connect with a mentor.	http://www.nbha.org/
Global	Fragility Fracture Network	FFN website: A comprehensive suite of resources relating to perioperative care of fragility fracture patients, surgical treatment, rehabilitation, secondary prevention and changing health-care policy.	http://fragilityfracture network.org/
Global	International Osteoporosis Foundation	IOF Capture the Fracture Program website: A comprehensive web-based resource which includes an FLS Implementation Toolkit, slide kits, a webinar series, opportunities for mentorship and the IOF capture the Fracture Best Practice Framework.	http://capturethe fracture.org/

Index

Note: 'Page numbers followed by "f" indicate figures, "t" indicate tables'.

A

Alberta Provincial Fracture Liaison Services, 87–88
American Orthopaedic Association (AOA), 156
American Society for Bone Mineral Research (ASBMR), 123, 155
Amiens University Hospital, 41
Ankle fractures, 14–15
Antiresorptive medication, osteoporotic refracture, 22–23
Arab health-care systems, 119
Asymptomatic vertebral fractures, 14
Atraumatic fractures, 10–11
Australia
  clinical standards, 173–174
  Fracture Liaison Services (FLS)
    activity-based funding, 70
    acute care costs, 67–68
    Australian and New Zealand Hip Fracture Registry, 70–71, 72f, 73t
    clinical outcome, 68
    dietary and lifestyle education, 69
    follow-up, 68
    GP-based care, 69
    hospital-based fracture liaison service, 69
    human and financial costs, 64
    long-term conditions, 67–68
    National Action Plan, 67–68
    National Service Improvement Framework, 67–68
    NSW Agency for Clinical Innovation, 69
    randomised controlled trial, 68
    Re-Fracture Prevention Service Directory, 69–70
    SOS Fracture Alliance, 68
  Fracture Liaison Services/secondary fracture prevention program (FLS/SFPP) implementation initiatives, 173
  fracture registries, 174
  health care, 63–64
  Medicare, 63–64
  Pharmaceutical Benefits Scheme, 63–64
  population, 63–64
  state/territory, 63–64
  Westminster system, 63–64
Australian and New Zealand Hip Fracture Registry (ANZHFR), 70–71, 72f, 73t, 174
Australian Commission on Safety and Quality in Health Care, 70–71
Australian Institute of Health and Welfare, 64
Austrian Nationwide Study, 22–23

B

Beirut Fracture Liaison Services (FLS) model, 119
Best Practice Framework (BPF), 133
Best Practice Tariff (BPT), 149
Bisphosphonate therapy, 22–23, 39
Bone and Joint Decade Global Network Conference, 156
Bone and Joint Japan poster, 112, 113f
BoneCare 2020, 66, 175
Bone protection medication, 70–71, 72f

C

Canada
  clinical standards, 175
  Fracture Liaison Services/secondary fracture prevention program (FLS/SFPP) implementation initiatives, 174
  fracture registries, 175
  geographic distribution, 79
  health-care system, 80
  Osteoporosis Canada (OC). *See* Osteoporosis Canada (OC)
  population density, 79
Canada Health Transfer, 80
Canadian Association of Radiologists/Osteoporosis Canada (CAROC), 81
Canadian Multicentre Osteoporosis Study, 16–17, 20

Capture the Fracture campaign, 133
CAROC. *See* Canadian Association of Radiologists/Osteoporosis Canada (CAROC)
CECity cloud-based registry, 167
Centralised computerised database (CCRD), 127, 129
China Hip Fracture Registry, 4
Chronic Disease Management Program (CDMP), 124
Colles fractures, 14–15
Community-based exercise, 126–127
Concord Hospital Secondary Fracture Prevention (SFP) programme, 35, 35f, 39, 42–43
Concord Secondary Fracture Programme, 16

D

Danish registry study, 12–13, 22–23
Demographic characteristics and fracture information, 91–92, 92t
Distal forearm fractures, 14–15, 44
Dual-energy X-ray absorptiometry (DEXA) screening, 119
Dubbo Osteoporosis Epidemiology Study, 11–14, 16–17, 19–21

E

e-consult, 46
Educational intervention system, 58
Effective interventions in health and social care, 149–150, 178
Eldershield, 123–124
Electronic health care databases, 17
Electronic health record, 35
Electronic Medical Record Exchange (EMRX), 126
EMERGE, 126
Essential Elements of Fracture Liaison Services, 175

F

Falls assessment, 4
Federation of Australia, 63–64
FFN. *See* Fragility fracture network (FFN)
5IQ approach, 179
Fracture Clinic Screening Program, 174
Fracture Liaison Service (FLS), 2
  in Australia
    activity-based funding, 70
    acute care costs, 67–68
    Australian and New Zealand Hip Fracture Registry, 70–71, 72f, 73t
    clinical outcome, 68
    dietary and lifestyle education, 69
    follow-up, 68
    GP-based care, 69
    hospital-based fracture liaison service, 69
    human and financial costs, 64
    long-term conditions, 67–68
    National Action Plan, 67–68
    National Service Improvement Framework, 67–68
    NSW Agency for Clinical Innovation, 69
    randomised controlled trial, 68
    Re-Fracture Prevention Service Directory, 69–70
    SOS Fracture Alliance, 68
  in Canada. *See* Osteoporosis Canada (OC)
  fragility fractures, 119
  Japan, 110
  in New Zealand
    Australian and New Zealand Hip Fracture Registry, 70–71, 72f, 73t
    BoneCare 2020, 66
    falls and fracture prevention, 66–67
    set of indicators, 66
  in Taiwan. *See* Taiwan
  United Kingdom (UK) National Health Service. *See* United Kingdom (UK) National Health Service
  in United States
    demonstration project, 165, 165f
    FLS registry, 156–157
    at Geisinger Health System, 155
    Geisinger High-Risk Osteoporosis Clinic Programme. *See* Geisinger High-Risk Osteoporosis Clinic Programme
    Kaiser Permanente Healthy Bones Programme. *See* Kaiser Permanente Healthy Bones Programme
    National Bone Health Alliance (NBHA), 155–157
    Own the Bone, 156, 166–168
    postfracture care, 155
    quality reporting, 168–169
    registries/databases, 167–168
    return on investment calculator, 157
    Symposium, 156
Fracture risk assessment tools, 81
Fragility Fracture Network (FFN), 3–4, 119, 181–182
  Japan annual meeting, 111, 111f
  orthogeriatric approach, 110

policy change, 110
for secondary prevention, 110
vision and mission, 110
Fragility fractures
ageing population, 1
annual global incidence of, 1
Australia, 64
and bone health, 2, 97–98
Canada, 80
in central Auckland region, 2–3
education programs, 97
epidemiological studies, 2
Fracture Liaison Service/Secondary Fracture Prevention Program (FLS/SFPP), 4, 5t
Fragility Fracture Network, 3–4
Glasgow Fracture Liaison Service (FLS), 2
hip fracture, 2–4
in Lebanon
epidemiology, 117–118
evidence-based medicine, 118
Fracture Liaison Service (FLS), 119
Fragility Fracture Network, 119
International Osteoporosis Foundation (IOF), 119
LOPS promoting regular integration, 118
Ministry of Public Health (MoPH), 119–120
osteoporosis diagnosis and treatment, 117
patient education and rehabilitation, 117–118
population, 117
secondary prevention, 118
morbidity and mortality, 117
National Hip Fracture Database (NHFD), 3–4
New Zealand, 66
nonpharmacological strategies, 98
primary care providers (PCPs), 100
psychological and physical burden, 98
qualitative methods, 98
refractures, 2
systematic approach, 3, 3f
Framingham study, 13
FRAX tool, 81

G

Geisinger High-Risk Osteoporosis Clinic Programme
database, 167–168
GHP Projected Healthcare Savings, 161, 161f
Glucocorticoid-Induced Osteoporosis (GIOP) programme, 161
healthcare, 160–161
High-Risk Osteoporosis Clinic (HiROC) pathways. *See* High-Risk Osteoporosis Clinic (HiROC) pathways
metrics, 167–168
Osteoporosis Disease Management Programme, 161
primary care network, 161
Geisinger leadership, 164–165
Geisinger primary care physicians (G-PCP), 162–164, 162t, 164t
Genant semiquantitative technique, 134–135
Glasgow model, 43
Glucocorticoid-Induced Osteoporosis (GIOP) programme, 161
Guidelines for Fragility Fractures in Lebanon, 117–118

H

Healthcare Effectiveness Data and Information Set (HEDIS) measurement, 159–160
Healthy Ageing Strategy, 176
Healthy Bones Programme, 35
Helen Hayes Hospital, 52
HIDS. *See* Hospital Inpatient Discharge System (HIDS)
High-Risk Osteoporosis Clinic (HiROC) pathways
care gaps, 164
database coordinator, 167–168
disadvantages, 167–168
facilitation of care, 164
follow-up, 162
Geisinger leadership, 164–165
vs. Geisinger primary care physicians (G-PCP), 162–164, 162t, 164t
geographic sites, 162–163
inpatient HiROC, 161–163
Microsoft Access software database, 162–163
outpatient pathway, 161–163
registry, 162–163
short-term rehabilitation, 164
treatment rates, 163–164
Hip Fracture Care Clinical Care Standard, 70–71, 173–174
Hip fractures, 70–71, 72f, 73t
China Hip Fracture Registry, 4
clinical pathways, 109
Fragility Fracture Network, 3–4
incidence, 10–12, 11f
mortality rates, 124

Hip fractures *(Continued)*
mortality risk, 18–19
National Hip Fracture Database (NHFD), 3–4
National hip fracture registries, 3–4
in Niigata Prefecture, 112–115, 113f, 114t
posthip fracture mortality, 20
refractures, 2
secondary prevention, 3–4
subsequent fracture risk, 13
Hospital-based fracture liaison service, 69
Hospital Inpatient Discharge System (HIDS), 126
HSDP Osteoporosis Management Program, 125

I

ICD-10 coding data, 41
International Osteoporosis Foundation (IOF), 5, 111, 119, 123, 133
International Osteoporosis Foundation (IOF) Capture the Fracture Best Practice Framework, 112–115, 180–181

J

Japan
Fracture Liaison Service (FLS) model, 110
Fragility Fracture Network (FFN), 110–111, 111f
health-care system, 109
hip fracture clinical pathways, 109
hip fracture in Niigata Prefecture, 112–115, 113f, 114t
Japan National Hip Fracture Database (JNHFD), 111–112, 112f–113f
osteoporosis liaison service (OLS) model, 110
Japanese Orthopaedic Association (JOA), 109
Japan National Hip Fracture Database (JNHFD), 111–112, 112f–113f
John Hunter Hospital SFP programme, 39–41

K

Kaiser Permanente Healthy Bones Programme
Care Managers, 158–160
case finding, 158–159
continuous quality improvement (CQI) culture, 158
effectiveness, 158–159, 159f–160f
electronic health-care technology, 158
Healthcare Effectiveness Data and Information Set (HEDIS) measurement, 159–160
Kaiser Foundation Health Plans, 157–158
Kaiser Foundation Hospitals, 157–158
midlevel providers, 158–159
Plan-Do-Study-Act (PDSA), 158
planning and goal setting, 158–159
regional Permanente medical groups, 157–158
Simple Model, 158
work lists, 158
Kaiser Permanente's custom-programmed registry, 167
Kaiser Permanente Secondary Fracture Prevention (SFP) programme, 42
Kaplan–Meier survival analysis, 129
Key performance indicators (KPIs), 125

L

Lebanese Osteoporosis Prevention Society, 117–118
Lifestyle and Fitness Enhancement Centre, 126
Life tables analysis, 15, 18–19
Local Health Integration Networks (LHINs), 83
Lower limb fractures, 14–15
Low-trauma fracture, 63, 65–66, 69, 71

M

Manitoba DXA health-care database, 15
Markov model, 96
MAXCARE, 126
McGuire Veterans Affairs Medical Center, 44–46
Mean Medication Possession Ratio (MPR), 129
Medicare, 63–64
Medifund, 123–124
Medisave, 123–124
Medishield, 123–124
Metabolic Bone Clinic (MBC), 93–94
Microsoft Access software database, 162–163
Minimum Standards for the Management of Hip Fracture in the Older Person, 173–174
Ministry of Public Health (MoPH), 119–120
Model of care, 34
Multivariate Cox regression, 42

N

National Action Plan, 67–68
National Bone Health Alliance (NBHA), 155–157
National Campaign for Osteoporosis Awareness, 119–120

National Health Insurance (NHI), 136
National Hip Fracture Database (NHFD), 3–4, 43, 146–147
National hip fracture registries, 2–3
National Institute for Health and Care Excellence (NICE), 148, 178–179
National Osteoporosis Society, 178
National Service Improvement Framework, 67–68
National Taiwan University Hospital (NTUH) Healthcare System, 134
antiresorptive medications, 136
asymptomatic vertebral compression fractures, 134–135
Bei-Hu branch, 134
case coordinators, 134
flow chart, 134, 135f
follow-up period, 136
Genant semiquantitative technique, 134–135
hip fracture cases, 134–135
for latent vertebral compression fractures, 136
medication review, 136
multifaceted evaluations, 136
National Health Insurance (NHI), 136
symptomatic vertebral compression fractures, 134–135
New Zealand
clinical standards, 176–177
Fracture Liaison Services (FLS)
Australian and New Zealand Hip Fracture Registry, 70–71, 72f, 73t
BoneCare 2020, 66
falls and fracture prevention, 66–67
set of indicators, 66
Fracture Liaison Services/secondary fracture prevention program (FLS/SFPP) implementation initiatives
BoneCare 2020, 175
District Health Board Alliances, 175–176
falls and fractures outcomes framework, 176, 177f
Fracture Liaison Service (FLS) Business plan template, 176
Fracture Liaison Service (FLS) Resource pack, 176
Fracture Liaison Service (FLS) Status summary, 176
Fracture Liaison Service (FLS) Step-by-step guide, 176
fracture risk assessment tools, 175
Generic Fracture Liaison Nurse job description, 176
Healthy Ageing Strategy, 176
investment, 176
outcomes and quality of care, 175
stakeholder organisations, 175
fracture registries, 177–178
health-care system, 64
The Ministry of Health, 64
Pharmaceutical Management Agency of New Zealand (PHARMAC), 64
Westminster system, 64
New Zealand Accident Compensation Corporation, 66
New Zealand Health Quality and Safety Commission, 70–71
NHFD. *See* National Hip Fracture Database (NHFD)
NHI. *See* National Health Insurance (NHI)
NICE. *See* National Institute for Health and Care Excellence (NICE)
Niigata Rehabilitation Hospital, 112–115
Nonhip nonvertebral fracture (NHNV), 9, 11
incidence, 11
mortality risk, 20
subsequent fracture, 14–15
Nonosteoporotic traumatic vertebral fractures, 10–11
Nonspine fracture, 13–14
Norwegian and United Kingdom (UK) study, 17
Nova Scotia Provincial Fracture Liaison Services, 86–87
NTUH Healthcare System. *See* National Taiwan University Hospital (NTUH) Healthcare System

O

OC. *See* Osteoporosis Canada (OC)
OLS model. *See* Osteoporosis liaison service (OLS) model
'1i' model, 4
One-Stop Notebook, 112–115, 113f
On-site dual-energy X-ray absorptiometry (DEXA) scanning, 33, 41
Ontario Osteoporosis Strategy (OOS), 83, 174
Ontario Provincial Fracture Liaison Services
Fracture Screening and Prevention Program (FSPP)
bone mineral density (BMD) 'fast track' design, 84–85
bone mineral density (BMD) testing, 85
clinical and diagnostic supports, 83–84
data collection and program evaluation, 85–86
education, 84

Ontario Provincial Fracture Liaison Services (*Continued*)
limitations, 86
patient data security, 86
patient identification, 84
pharmacotherapy, 85
resource allocation, 86
Local Health Integration Networks (LHINs), 83
Ontario Health Insurance Plan, 83
Ontario Osteoporosis Strategy (OOS), 83
The Osteoporosis Action Plan, 83
OPTIMAL. *See* Osteoporosis Patient Targeted and Integrated Management for Active Living (OPTIMAL)
Oral bisphosphonate treatment, 47
Osteoporosis assessment, 4
Osteoporosis Canada (OC)
bone mineral density (BMD) testing, 80
Canadian Association of Radiologists/ Osteoporosis Canada (CAROC), 81
clinical practice guidelines, 80
Fracture Liaison Services (FLS)
adaptability to local conditions, 101
Alberta Provincial Fracture Liaison Services, 87–88
components, 82
consultation service, 82
continuous quality improvement (CQI), 82
coordinator/nurse role, 101
cost-effectiveness, 96–97
data collection, 101
description, 82, 82t
evidence-based recommendations, 97
gender equality, 101
independent programmes, 88–96, 89t–90t
iterative improvement processes, 100–101
national audit, 82
Nova Scotia Provincial Fracture Liaison Services, 86–87
Ontario Provincial Fracture Liaison Services. *See* Ontario Provincial Fracture Liaison Services
patient's knowledge, perceptions and attitudes, 102
quality standards, 82
registry, 82
at St. Michael's Hospital, Toronto. *See* St. Michael's Hospital, Fracture Liaison Services (FLS)
toolkit, 82
FRAX tool, 81
pharmacotherapy, 81
Osteoporosis care management, 134
Osteoporosis Disease Management Programme, 161
Osteoporosis liaison service (OLS) model, 110
Osteoporosis New Zealand, 66
Osteoporosis Patient Targeted and Integrated Management for Active Living (OPTIMAL)
block funding, 130–131
care coordinator, 129
centralised computerised data base (CCRD), 127, 129
clinician champion, 129
components, 127, 128f
delivery processes, 129
electronic medical record, 126
Electronic Medical Record Exchange (EMRX), 126
electronic patient records management system, 126
exercise programmes, 126–127
falls risk assessment, 126–127
follow-up care, 126–127
fracture case records, 126–127
Hospital Inpatient Discharge System (HIDS), 126
HSDP Osteoporosis Management Program, 125
inclusion criteria, 127, 127t
Kaplan–Meier survival analysis, 129
key performance indicators (KPIs), 125
laboratory work, 126–127
Mean Medication Possession Ratio (MPR), 129
medications, 125
operational characteristics, 127
with orthogeriatric and/or hip fracture care pathways, 130–131
Osteoporosis Prevention and Treatment Initiative, 126–127
patient-level challenges, 130
pharmacy patient purchase records system, 126
in polyclinics, 124–127
problems and limitations, 129–130
in public acute general hospitals, 124–125
system-level challenges, 130
Vertebral Fracture Assessment, 130
workflow, 127, 128f
Osteoporosis Prevention and Treatment Initiative, 126–127
Osteoporotic fracture triad, 12
Osteoporotic refracture

antiresorptive medication, 22–23
definition, 9
direct and indirect costs, 9
incidence
in Dubbo population, 9–12, 10f
hip fracture, 10, 11f
nonhip nonvertebral fracture (NHNV), 11
vertebral fractures, 10–11
in industrialised countries, 9
initial fracture risk, 11–12
mortality risk, 15–16, 21
epidemiological studies, 18
hip fracture, 18–19
nonhip nonvertebral fracture (NHNV), 20
risk factors, 18
vertebral fracture, 19
postfracture mortality
posthip fracture mortality, 20
secular trends, 21
symptomatic vertebral fractures, 21
risk factors, 9
subsequent fracture
bone mineral density (BMD), 16–17
Danish hip fracture study, 15
initial hip fracture, 13
initial nonhip nonvertebral fracture (NHNV), 14–15
initial vertebral fracture, 13–14
long-term risk, 15–16
Manitoba DXA health-care database, 15
minor fractures, 15
risk, 12–18
secular trends, 17–18
timing of, 15–16
OTAGO, 126–127
Own the Bone, 166–168

P

Pharmaceutical Benefits Scheme, 63–64
Pharmaceutical Management Agency of New Zealand (PHARMAC), 64
Physiotherapy exercise, 126–127
Polyclinics, 124
Post Fracture Osteoporosis (PFO) Clinic, 92–93
Primary Health Networks, 5
Princess Margaret Hospital, 46
Programmes of care, 69

Q

Quality and Outcomes Framework (QOF), 149
Quality Standards for Fracture Liaison Services, 175

R

Re-Fracture Prevention Service Directory, 69–70
Regional Cooperation of Clinical Pathways for Hip Fracture, 109
Reykjavik study, 15
Rotterdam study, 10–11
Royal College of Physicians Falls and Fragility Fracture Audit Programme, 146–147, 148f
Royal Melbourne Hospital SFP programmes, 43

S

Secondary Fracture Prevention Program (SFPP), 4
degree of intervention, 33
design and clinical efficacy, 33
heterogeneity, 34
intervention intensity and intervention outcomes, 57, 57t
intervention types, 33, 34t
models of care, 34t
on-site DXA scanning, 33
outcome measures, 34
type A model interventions
adherence, 39–42
assessments and investigations, 35
bone mineral density (BMD) testing, 36, 38f
care managers, 35
clinical risk factors, 35
closed system approach, 35
Concord Hospital SFP programme, 35, 35f
coordinated model of care, 35
cost-effectiveness, 36, 43
electronic health record, 35
metaanalysis, 36, 37t
pharmacological treatment, 36–39, 40f
refractures, 42
type B model interventions
adherence, 47
bone mineral density (BMD) testing, 44–46, 45t
cost-effectiveness, 47–48, 49f–50f
financial incentives, 43
Glasgow model, 43
metaanalysis, 44, 45t
National Hip Fracture Database (NHFD), 43
type C model interventions

Secondary Fracture Prevention Program *(Continued)*
bone mineral density (BMD) testing, 48–53, 54f–55f
'face-to-face' interviews, 48
meta-analysis, 48, 51t
type D model interventions, 53–57, 56t
up-to-date metaanalysis, 57
SFPP. *See* Secondary Fracture Prevention Program (SFPP)
Simple Model, 158
Singapore
Chronic Disease Management Program (CDMP), 124
Eldershield, 123–124
health-care environment, 123
medical care, 124
Medifund, 123–124
Medisave, 123–124
Medishield, 123–124
Osteoporosis Patient Targeted and Integrated Management for Active Living (OPTIMAL). *See* Osteoporosis Patient Targeted and Integrated Management for Active Living (OPTIMAL)
osteoporotic fractures, 124
out-of-pocket charges, 123–124
polyclinics, 124
population, 123–124
restructured hospitals, 124
St. Michael's Hospital, Fracture Liaison Services (FLS)
baseline quality assurance questionnaire, 95
bone health assessment
inpatients, 94–95
outpatients, 93–94
pharmacotherapy, 94–95, 99
bone mineral density (BMD) testing, 93, 99
coordinator role, 88–91, 91f
data collection and analysis, 88–91, 95
education to patients, 92–93, 93f, 101
fragility fracture and bone health, 97–98
iterative program modifications, 88–91
Musculoskeletal Health and Outcomes Group, 97
orthopaedic research fund, 88–91
patient identification, 91–92, 92t
strengths and limitations, 95–96
Stop Osteoporosis Secondary (SOS) Fracture Alliance, 68, 173
Study of Osteoporotic Fractures, 10–11
St. Vincent's Hospital, 39–41
Sunrise Clinical Manager, 126

T

Taiwan
annual incidence of hip fracture, 133
bone mineral density (BMD) testing, 133
Fracture Liaison Service (FLS)
Best Practice Framework (BPF) standards, 138–141, 141f
challenges, 143
characteristics of, 137–138, 139t–140t
DXA scanners, 138
E-Da Hospital (EDAH), 138
established sites, 137, 137f
informatics system, 138
International Osteoporosis Foundation–accredited sites, 137, 137f
long-term management schedule, 142
one-year mortality, 142, 142f
patient evaluation, 138–141
mortality, 133
National Taiwan University Hospital (NTUH) Healthcare System. *See* National Taiwan University Hospital (NTUH) Healthcare System
osteoporosis care management, 134
Taiwanese Osteoporosis Association (TAO), 136
The Agency for Clinical Innovation's Musculoskeletal Network, 69
The Australian and New Zealand Bone and Mineral Society, 66
The European Prospective Osteoporosis study, 10–11
The Kaiser Permanente Group in Southern California, 35
The Ministry of Health, 64
The Osteoporosis Action Plan, 83
'3i' model, 4
Toronto-based Secondary Fracture Prevention programme, 43
Trans-Tasman publications, 66
Tromsø study, 12–13
'2i' model, 4

U

United Kingdom (UK) General Practice Research Database, 13
United Kingdom (UK) National Health Service
challenges, 150–151
clinical and economic framework, 145
clinical standards, 178–179
Fracture Liaison Service (FLS) concept, 146
benefits calculator, 149–150
Best Practice Tariff (BPT), 149
cost calculator, 147–148

Database Audit, 146–147
Department of Health systematic approach, 146, 147f
financial incentives, 149
national FLS champions, 149–150
National Hip Fracture Database (NHFD), 146–147
National Institute for Health and Care Excellence (NICE), 148
online Fracture Liaison Service (FLS) benefits, 147–148
Quality and Outcomes Framework (QOF), 149
quality standards, 147–148
Royal College of Physicians Falls and Fragility Fracture Audit Programme, 146–147, 148f
stakeholders, 147–148
UK National Osteoporosis Society (NOS), 147–148
Fracture Liaison Services/secondary fracture prevention program (FLS/SFPP) implementation initiatives, 178
fracture registries, 179–180
independent general practitioners (GP), 145
nonphysician-delivered service, 146
preceptorships, 146
principles, 145
United Kingdom (UK) National Osteoporosis Society (NOS), 147–148
United States of America
clinical standards, 180
Fracture Liaison Service (FLS). *See* Fracture Liaison Service (FLS)
Fracture Prevention CENTRAL website, 180
fracture registries, 180
public–private partnership, 180
University Hospital of Saint Etienne, 41–42
University of Wisconsin Medical Foundation, 44

V

Vancouver-based prospective controlled study, 52
Vertebral deformities, 14, 19
Vertebral fractures
incidence, 10–11
mortality risk, 19
postfracture mortality, 21
subsequent fracture, 13–14
Veterans Affairs Medical Centers, 46

W

West Glasgow Secondary Fracture Prevention (SFP) programme, 47
World Osteoporosis Day Report, 1
Wrist fractures, 52–53

Z

'Zero i' model, 4
Zoledronic acid study, 22

Printed in the United States
By Bookmasters